A TEXT BOOK OF

ANIMAL DIVERSITY-II

Paper V

&

BIOCHEMISTRY

Paper VI

FOR

B.Sc. Part II Zoology [DSC 15C and DSC 16C]: Semester - III

As Per New Revised CBCS Syllabus of Shivaji University, Kolhapur with effect from June 2019

Dr. Jadhav Bapurao Vishnu
M. Sc. Ph. D
Assistant Professor,
Department of Zoology,
Balasaheb Desai College, Patan,
Dist. Satara.

Dr. Atigre Rajaram Hindurao
M. Sc., M. Phil, Ph. D., SET
Assistant Professor & Head,
Department of Zoology,
Shri Vijaysinha Yadav Arts and Science
College, Peth Vadgaon, Dist. Kolhapur.

Dr. Chougale Tanaji Mahadev
M. Sc. Ph. D, SET
Assistant Professor
Department of Zoology,
Bhogawati Mahavidyalaya, Kurukali,
Tal. Karveer, Dist. Kolhapur.

Mr. Mane Abhijit Bhagwan
M. Sc.
Assistant Professor
Department of Zoology,
Arts Science nerce College,
i), Dist. Sangli

N4820

Animal Diversity-II (P-V) & Biochemistry (P-VI)   ISBN : 978-93-89533-57-6

First Edition : **October 2019**

© : **Authors**

Published By:
NIRALI PRAKASHAN
Abhyudaya Pragati, 1312, Shivaji Nagar
Off J.M. Road, PUNE – 411005
Tel - (020) 25512336/37/39, Fax - (020) 25511379
Email : niralipune@pragationline.com

✍ DISTRIBUTION CENTRES

PUNE

Nirali Prakashan : 119, Budhwar Peth, Jogeshwari Mandir Lane,
Pune 411002, Maharashtra. Tel : (020) 2445 2044, 66022708,
Fax : (020) 2445 1538, Email: bookorder@pragationline.com,
niralilocal@pragationline.com

Nirali Prakashan : S. No. 28/27, Dhyari, Near Pari Company, Pune 411041
Tel : (020) 24690204 Fax : (020) 24690316
Email : dhyari@pragationline.com,
bookorder@pragationline.com

MUMBAI

Nirali Prakashan : 385, S.V.P. Road, Rasdhara Co-op. Hsg. Society Ltd.,
Girgaum, Mumbai 400004, Maharashtra
Tel : (022) 2385 6339 / 2386 9976, Fax : (022) 2386 9976
Email : niralimumbai@pragationline.com

✍ DISTRIBUTION BRANCHES

JALGAON

Nirali Prakashan : 34, V. V. Golani Market, Navi Peth, Jalgaon 425001,
Maharashtra, Tel : (0257) 222 0395,
Mob : 94234 91860

KOLHAPUR

Nirali Prakashan : New Mahadvar Road, Kedar Plaza, 1st Floor Opp. IDBI Bank
Kolhapur 416 012, Maharashtra. Mob : 9850046155

NAGPUR

Pratibha Book Distributors : Above Maratha Mandir, Shop No. 3, First Floor,
Rani Jhanshi Square, Sitabuldi,
Nagpur 440012, Maharashtra Tel : (0712) 254 7129

DELHI

Nirali Prakashan : 4593/15, Basement, Aggarwal Lane, Ansari Road,
Daryaganj, Near Times of India Building, New Delhi 110002
Mob : 08505972553

BANGALURU

Nirali Prakashan : Maitri Ground Floor, Jaya Apartments, No. 99, 6th Cross,
6th Main, Malleswaram, Bangaluru 560 003, Karnataka
Mob : +91 9449043034
Email: niralibangalore@pragationline.com

www.pragationline.com   **info@pragationline.com**

PREFACE

We take this opportunity to express our sense of gratitude to all our esteemed teachers from the Department of Zoology, Shivaji University Kolhapur, for their continuous efforts on updating the knowledge of subject from time to time.

It is a great privilege to present this book to teachers and students of B.Sc. - II Zoology, Shivaji University, Kolhapur. The book has been prepared as per CBCS syllabus given by UGC and implemented by the University from June 2019. The book is divided into two parts for making it handy. The first part deals with Paper V, Animal Diversity - II & second part deals with Paper VI, Biochemistry including metabolism of biomolecules. Total twelve chapters are divided into several heading and subheadings for the better understanding and memorizing the content. The language used in the book is simple and lucid. Clear printing, neat labelled diagrams, self-explanatory charts and illustrations are important features of this text book.

In writing the book references of various authors has been used. The authors are extremely grateful to them. The reference materials available on internet is also referred for the context enrichment of this book.

The authors are happy to express their heartfelt thanks to all those who helped, encouraged and guided us in writing of this book. The authors are aware of their limitations hence, creative and fruitful suggestions in making this book as a perfect study material for the students will be heartily welcomed.

We are extremely grateful to Nirali Prakashan for their timely assistance and for publishing this book in an attractive manner.

Authors

✩✩✩

PREFACE

SYLLABUS

Paper V : ANIMAL DIVERISTY-II

Unit I

1. Protochordates: (4 hrs.)

General characters and Classification of Protochordata.

2. Agnatha: (4 hrs.)

General characters of Agnatha and Classification of cyclostomes up to classes.

3. Pisces: (4 hrs.)

General characters and Classification up to orders; Respiration in Fishes.

4. Amphibia: (4 hrs.)

General features and Classification up to orders; Parental care.

Unit II

5. Reptiles: (4 hrs.)

General characters and Classification up to orders; Venomous and non-venomous snakes, Biting mechanism in snakes.

6. Aves: (5 hrs.)

General characters and Classification up to orders; Digestive and Respiratory systems.

7. Mammals: (5 hrs.)

General characters and Classification up to orders; Circulatory of mammals.

Paper VI : BIOCHEMISTRY

Unit I

1. Nucleic acids: DNA and RNA (7 hrs.)

Structure and types of RNA .DNA- Secondary structure of Watson and Crick. Forms of DNA

2. Carbohydrate Metabolism: (8 hrs.)

Glycolysis, Krebs Cycle, Pentose phosphate pathway, Gluconeogenesis, Glycogenolysis, Review of electron transport chain.

Unit II

3. Lipid Metabolism: (5 hrs.)

Biosynthesis and β oxidation of fatty acids.

4. Protein metabolism: (5 hrs.)

Transamination, Deamination and Urea Cycle.

5. Enzymes: (5 hrs.)

Introduction- classification and nomencelature. Mechanism of action, Enzyme Kinetics, Inhibition and Regulation. Isoenzymes, Co-enzymes and Co-factors.

☆☆☆

CONTENTS

☆☆☆

INTRODUCTION

Phylum Chordata

This phylum is probably the most notable phylum, as all lower chordates, Pisces, Amphibians, Reptiles Birds and Mammals including human beings are grouped under this phylum. The most distinguishing character that all animals belonging to this phylum is the presence of notochord.

Chordates show four distinguishing features, at different stages in their life. They are:

(i) Notochord : It is a longitudinal rod that is made of cartilage and runs between the nerve cord and the digestive tract. Its main function is to support the nerve cord. In Vertebrate animals, the notochord is replaced by vertebral column.

(ii) Dorsal Nerve Cord : This is a bundle of nerve fibres present in the dorsal part of body which connects the brain to the muscles and other organs. The nerve cord is single and hollow.

(iii) Post-anal tail : This is an extension of the body beyond the anus. In some chordates, the tail has skeletal muscles, which help in locomotion.

(iv) Pharyngeal slits : They are the openings which connect the mouth and the throat. These openings allow the entry of water through the mouth, without entering the digestive system. In higher vertebrates they are present only in larval stages.

The other characters of Chordates are :

(i) The habitat of these animals is widespread. So we can find them in the marine environment, fresh waters as well as terrestrial environments.

(ii) The body of all chordates is triploblastic and bilaterally symmetrical.

(iii) Chordates are coelomates and show an organ system level of organization.

(iv) They possess exoskeleton in the form of scales, scutes, plates, feathers, hairs, etc.

(v) Respiration is mostly by true gills or lungs.

(vi) The heart is present in ventral part of body, with a closed circulatory system and red coloured blood.

(1)

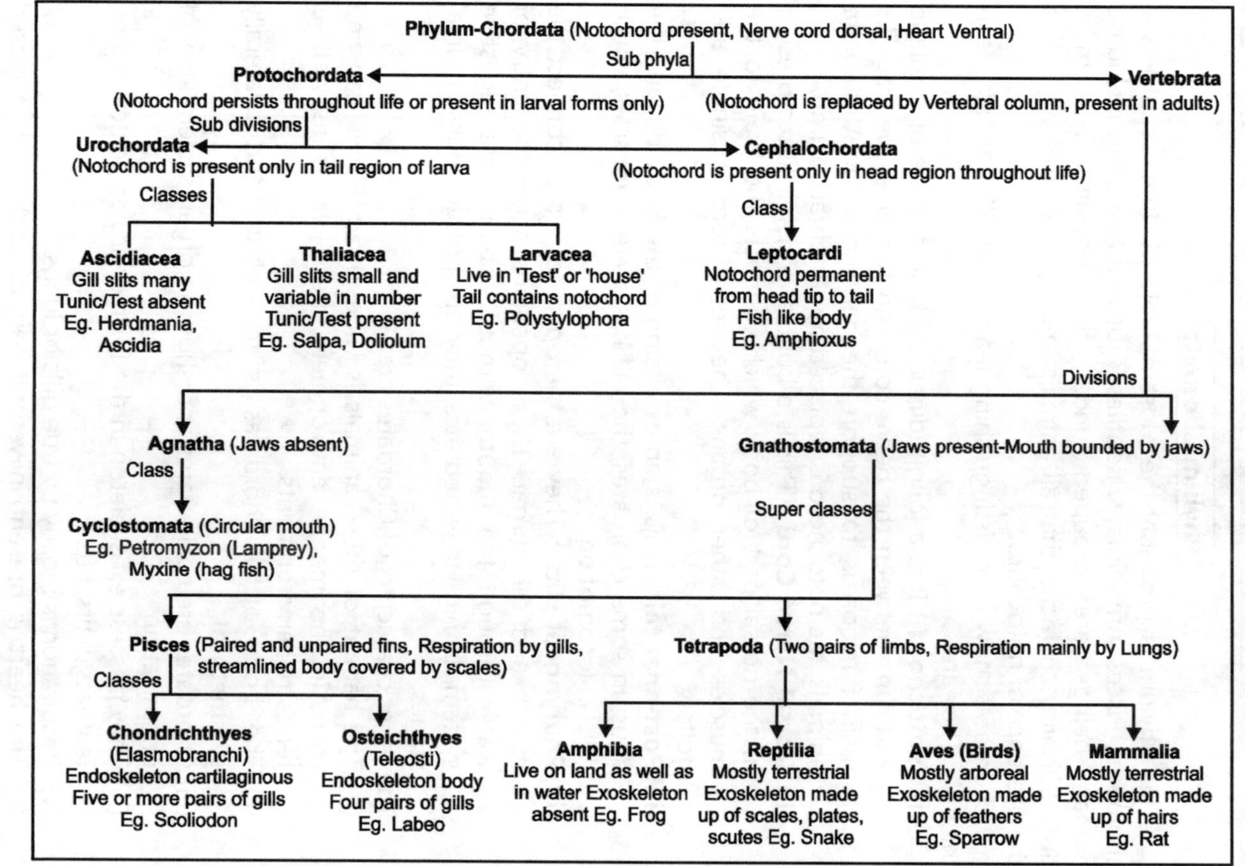
Phylum-Chordata (Notochord present, Nerve cord dorsal, Heart Ventral)
Sub phyla
Protochordata
(Notochord persists throughout life or present in larval forms only)
Vertebrata
(Notochord is replaced by Vertebral column, present in adults)
Sub divisions
Urochordata
(Notochord is present only in tail region of larva
Cephalochordata
(Notochord is present only in head region throughout life)
Classes
Class
Ascidiacea
Gill slits many
Tunic/Test absent
Eg. Herdmania,
Ascidia
Thaliacea
Gill slits small and
variable in number
Tunic/Test present
Eg. Salpa, Doliolum
Larvacea
Live in 'Test' or 'house'
Tail contains notochord
Eg. Polystylophora
Leptocardi
Notochord permanent
from head tip to tail
Fish like body
Eg. Amphioxus
Divisions
Agnatha (Jaws absent)
Class
Gnathostomata (Jaws present-Mouth bounded by jaws)
Super classes
Cyclostomata (Circular mouth)
Eg. Petromyzon (Lamprey),
Myxine (hag fish)
Pisces (Paired and unpaired fins, Respiration by gills, streamlined body covered by scales)
Classes
Tetrapoda (Two pairs of limbs, Respiration mainly by Lungs)
Chondrichthyes
(Elasmobranchi)
Endoskeleton cartilaginous
Five or more pairs of gills
Eg. Scoliodon
Osteichthyes
(Teleosti)
Endoskeleton body
Four pairs of gills
Eg. Labeo
Amphibia
Live on land as well as
in water Exoskeleton
absent Eg. Frog
Reptilia
Mostly terrestrial
Exoskeleton made
up of scales, plates,
scutes Eg. Snake
Aves (Birds)
Mostly arboreal
Exoskeleton made
up of feathers
Eg. Sparrow
Mammalia
Mostly terrestrial
Exoskeleton made
up of hairs
Eg. Rat

☆☆☆

1

CHAPTER

PROTOCHORDATES

1.1 INTRODUCTION

Protochordates are an informal category of animals (i.e. not a proper taxonomic group), named mainly for convenience to describe invertebrate animals that are closely related to vertebrates. Protochordates (Gr., protos = first, chorde = cord) or Protochordates are commonly called **lower chordates**. They lack a head and a cranium, so they are also known as **Acraniata**. Protochordates consists of three sub-phyla based on the property of notochord. They are **Hemichordata, Urochordata** and **Cephalochordata**. The division is mainly based on the notochord found in three subphyla.

1.2 HEMICHORDATA

Hemichordates where previously known as Enteropneusta a name given by Gegenbaur (1870) due to the presence of gill, gill-slits in *Balanoglossus clavigerus*. Bateson (1885) suggested the name Hemichordata instead of Enteropneusta, as he noticed many features of structures and development which these animals resemble with that of lower chordates like Cephalochordates and Urochordates. Although the name has been widely accepted, but its systemic position remained controversial due to presence of short stomodaeal diverticulum or stomochord to the proboscis. Hence, it is considered as separate phylum of Invertebrates by Hyman (1959).

General characteristic features:

- All hemichordates are marine. Some are solitary and slow moving, others are sedentary and colonial mostly tubicolous, soft and fragile animals.
- Body is divisible into proboscis, collar and trunk.
- They are vermiform, unsegemented, bilaterally symmetrical and triploblastic.

- The appendages are absent. In some the collar may bear arms with tentacles.
- Body wall with single layered epidermis. Dermis is totally absent.
- Coelom is enterocoelous divisible into protocoel or proboscis coelom, mesocoel or collar coelom and metacoel or trunk coelom.
- Buccal diverticulum, wrongly named as notochord, is present.
- Alimentary canal is complete, in the form of a straight or 'U' shaped tube.
- Respiration by gill-slits, one to several pairs are present. One pair in Cephalodiscus, several in Balanoglossus and absent in Rhabdopleura.
- Circulatory system is open type with a central sinus, heart vesicle, dorsal and ventral vessels, sinuses and lateral vessels.
- Excretion is performed by the glomerulus present in proboscis coelom and connected with blood vessels.
- Nervous system is primitive comprising mainly of an intra-epidermal nerve plexus.
- Mode of reproduction is sexual but some of them also exhibit asexual reproduction.
- Development may be direct i.e. without larval stage or indirect i.e. with a Tornaria larva.
- Hemichordates are exclusively marine and all feed on micro-organisms and debris by ciliary mechanism. There are about 70 known species to the world.

Classification:

Phylum Hemichordata is divided into following classes:

(1) Entropneusta

(2) Pterobranchia

(3) Planctosphaeroidea and

(4) Graptolita (Extinct)

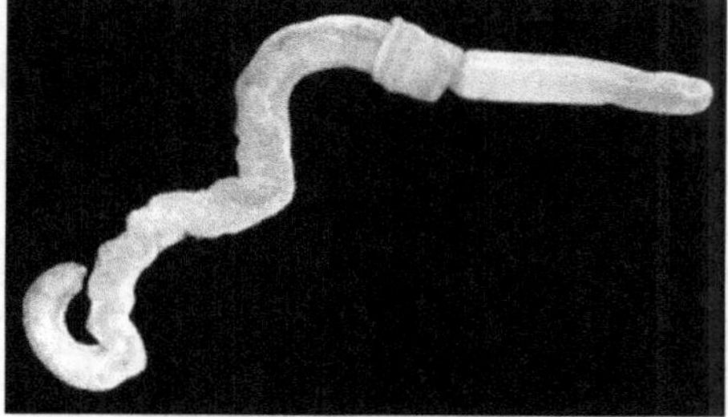

Fig 1.1 : Balanoglossus (Acorn worm)

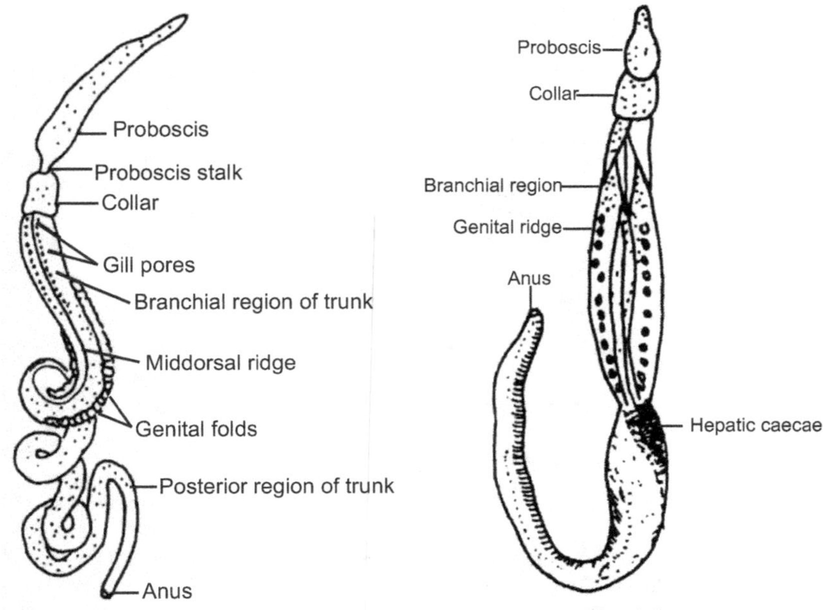

Fig. 1.2 : Saccoglossus

Fig. 1.3

(a) Class: Enteropneusta

1. They are free, solitary, burrowing animals, commonly known as 'Acorn' or 'Tongue worms'
2. Body is divisible into proboscis, collar and trunk. Proboscis is tapers anteriorly, collar without tentaculate arms.
3. Alimentary canal is straight tube having caecae in the intestinal region.
4. Several pairs of gill-slits.
5. Sexes are separate, gonads are numerous and sac-like.
6. Development is indirect through the Tornaria larval stage.

(b) Class: Pterobranchia

1. These are sedentary, colonial and tubicolous animals.
2. Body is very short and vase –like. Proboscis is shield shaped and the collar with hollow, ciliated arms bearing tentacles.
3. Gills-slits one pair or absent, if present it is never 'U'shaped.
4. The alimentary canal is 'U'shaped with anus near the mouth.
5. Bisexual or unisexual;one pair of gonad. Asexual reproduction by budding.
6. Development may be direct or indirect.

Class- Pterobranchia is divided into two orders:
Order (i) Rhabdopleurida
1. Collar with two tentaculated arms.

2. Gill-slits are absent.
3. Single gonad is found.
4. They are colonial; individuals of the colony are connected by a stolon.
   Example: Rhabdopleura.

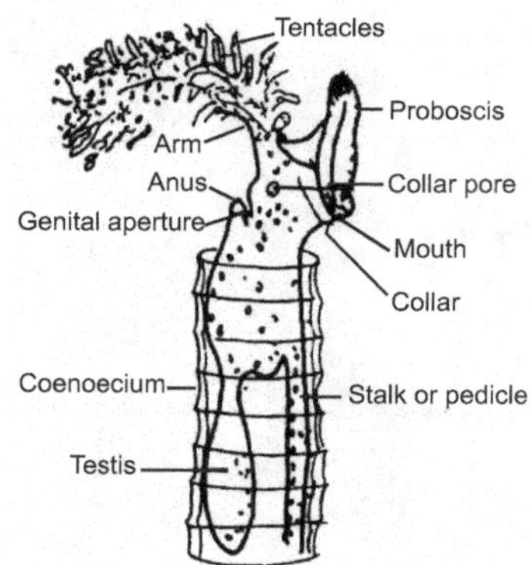

Fig. 1.4 : Rhabdopleura

Order (ii) Cephalodiscida

1. They are solitary but are enclosed in a common gelatinous case.
2. Collar with several tentaculated arms.
3. One pair of gill-slits is present.
4. One pair of gonads is present.
   Example: Cephalodiscus.

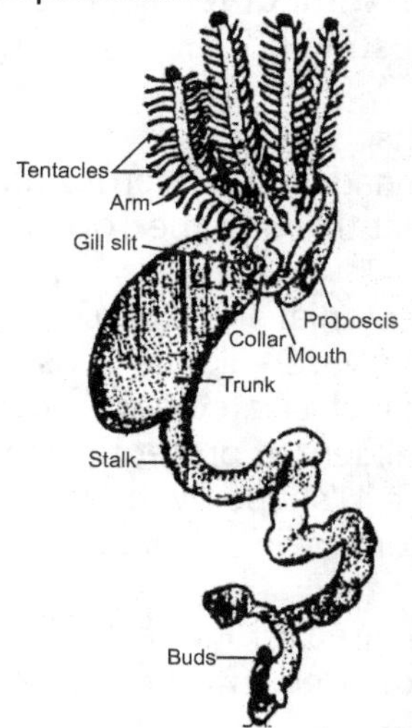

Fig. 1.5 : Cephalodiscus

(c) Class: Planctosphaeroidea

1. This class is represented by pelagic and transparent larvae.
2. Body is spherical with branched ciliated bands on the surface.
3. The alimentary canal is 'L' shaped.
4. A triangular protocoel opening outside by a hydropore. Paired mesocoel and metacoel are also present.

Example: Planctosphaera.

1.3 UROCHORDATA OR TUNICATA

The Urochordates are also known as Tunicates (L- tunica, an outer covering) or Asciadians (Gr. Askos, a leather bag). They were classified by Herdman on the basis of external morphology and ecological factors. Later on Lahille classified them on the basis of pharyngeal modifications. Garstang (1895) classified Urochordate on the basis of anatomical and embryological features. Perrier (1898) divided the urochordates into Enterogona, Pleurogona and Hypogona on the basis of their reproductive organs. The latest classification is given by Hartmeyer (1909 – 1911).There are about 2100 species of urochordates.

General characteristic features:

- They are **marine**, solitary or colonial, fixed or pelagic, found in all the seas. They   mostly sessile, filter–feeders.
- Body varies considerably in size, form and colour. Body is unsegmented and without tail. Body is enclosed in a leathery test or tunic sheath, composed of **tunicin (cellulose)** so called **tunicates**.
- They have two pores related to the atrial and branchial siphons.
- Coelom is a absent. However, an ectodermal lined atrial cavity surrounds the pharynx. The atrial cavity receives gonoducts, anus and gill- slits.
- The notochord occurs only in the tail of larva and disappears in the adults called **retrogressive metamorphosis**.
- The **nerve chord (neural tube)** is present in the larva, but is replaced by a single **dorsal ganglion** in the adult.
- The gill slits or stigma are numerous, persist in the adults and open into the atrium, instead of opening to the exterior. There are no true gills but are called **branchial basket**.

- Circulatory system is of open type. Blood consists of **Venadocytes**.
- Excretory system is lacking.
- Asexual reproduction occurs by **budding**.

Fig. 1.6 : Herdmania

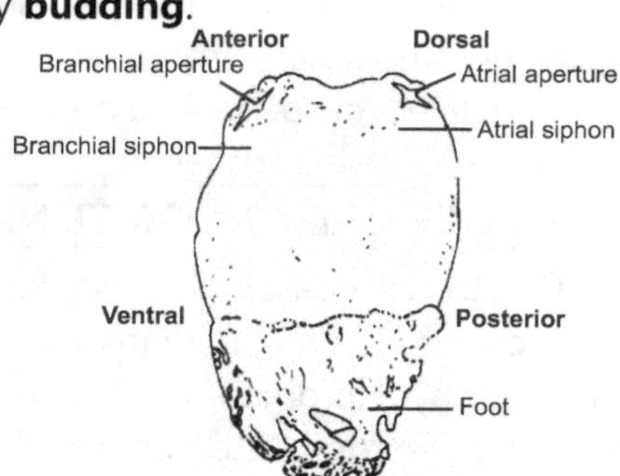

Fig. 1.7

Examples: Herdmania, Salpa, Doliolum etc.

Classification:

Sub-Phylum Urochordata is divided into three classes

1. Ascidiacea,
2. Thaliacea and
3. Larvacca

Class- 1. Ascidiacea

1. These are sessile, solitary or colonial tunicates with dorsal atriopore.
2. Pharynx is perforated, ciliated gill-slits open into the atrial cavity.
3. Sexes are united. Larva is free-swimming and highly developed. It undergoes retrogressive metamorphosis.
4. Adult is without notochord, nerve cord and tail.
5. Asexual reproduction by budding.

Class – Ascidiacea is divided into two orders: Enterogona and Pleurogona

Order (i) Enterogona

1. Body is sometimes divided into thorax and abdomen.
2. Neural gland is usually ventral to ganglion.

3. Gonad is unpaired and lodged in the intestinal loop or projecting behind it.

4. Larva with two sense organs, the otolith and cerebral eye or ocellus.

Order- Enterogona is divided into two sub-orders: Phlebobranchia and Aplousobranchia.

Sub-order (a) Phlebobranchia

1. Pharynx has an accessory, tubular longitudinal vessel.

2. The intestinal loop is usually along a side of the pharynx.

3. Epicardium in other cases is either sterile or absent.

4. Budding is rare.

    Examples: Ciona, Ascidia, Rhodosoma.

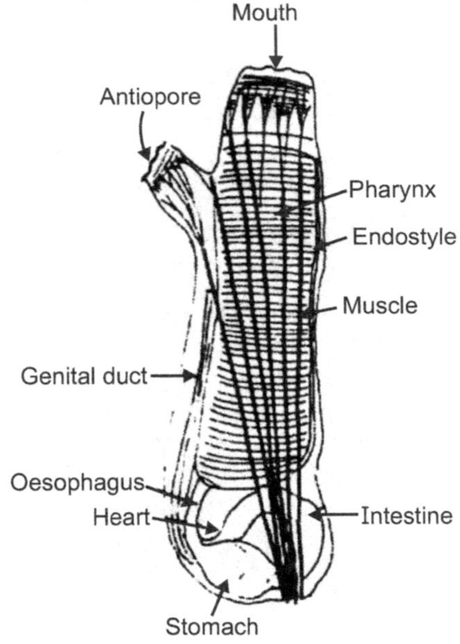

Fig. 1.8 : Ciona

Sub-order (b) Aplousobranchia

1. Pharynx with bars and longitudinal vessels but transverse ridges are present.

2. Body is elongated with distinct abdominal region.

3. The budding is common.

    Examples: Clavelina, Podoclavella.

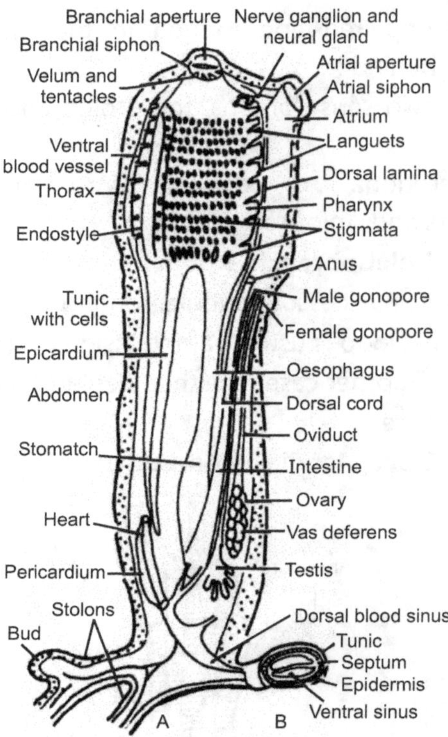

Fig. 1.9 : Clavellina (A) – Single zooid, (B) Stolon in T.S.

Order (ii) Pleurogona

1. Gonads are paired and lie in the atrial wall.
2. Neural gland is dorsal or lateral to the ganglion.
3. Larva has only one sense organ i.e. the otolith. The larval suckers are reinforced by adhesive papillae.
4. Budding is peribranchial and lateral.
   Example: Herdmania, Botryllus, Molgula.

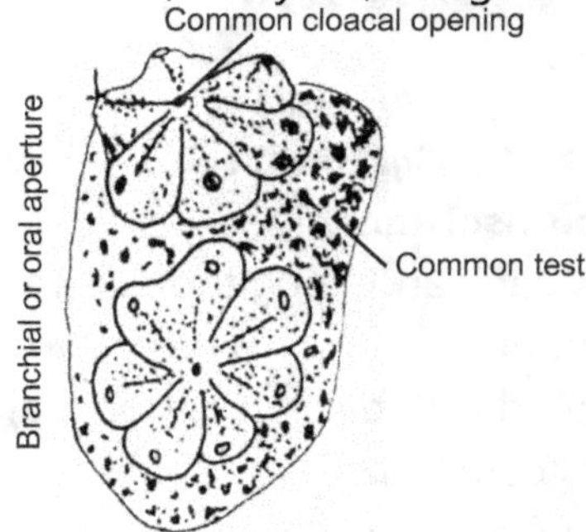

Fig. 1.10 : Botryllus

Class- 2 : Thaliacea

1. These are pelagic and hypopleustonic (animals floating below the surface of water) animals with transparent test.
2. Mouth and atriopore are situated at opposite ends.
3. Body is provided with numerous muscular bands which may encircle the body partially or wholly.
4. Life-cycle shows alteration of generation.
5. These forms reproduce sexually and asexually by complex budding.

   This class is divided into three orders: Pyrosomida, Doliolida and Salpida.

Order (i) Pyrosomida

1. Colonial forms hollow floating tubes, which open at one end only.
2. The cavity of the cylinder acts as a common cloaca.
3. Gill-slits are numerous and long.
4. Development is internal, larval stage is lacking.
5. The blastozooids are hermaphrodite and capable of budding. Oozoids are incompletely developmed.
6. Each individual has a cerebral eye and a subneural gland with ciliated funnel.
7. They are luminescent.

   Example: Pyrosoma.

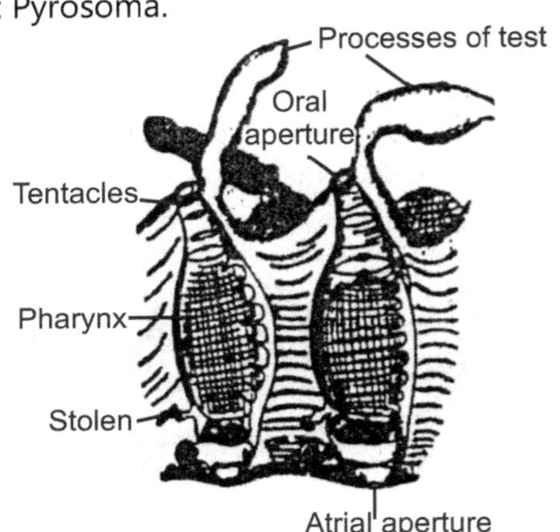

Fig. 1.11 : Pyrosoma

Order (ii) Doliolida

1. Body is charactically barrel shaped with thin test. Numerous muscle bands encircling the body.
2. Pharynx is provided with two rows of stigmata, without internal longitudinal bars.
3. The asexual phase is oozoid which reproduces by budding.
4. A tailed larva is present with notochord.
   Examples : Doliolum, Doliopsis.

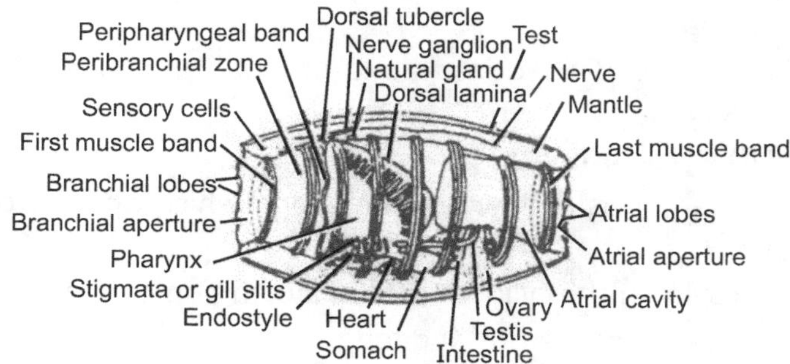

Fig. 1.12 : Structure of Doliolum

Order (iii) Salpida

1. Body is cylindrical or prism like.
2. The muscular bands are incomplete ventrally and attached to the test dorsally.
3. There is a single pair of very large lateral gill-slits.
4. The development is internal, inside the pouch of the body and the embryo is nourished by a placenta, Larval stage is absent.
5. In the life history alternation of generation is seen (as asexual phase or oozooid and a sexual phase or gonozooid).
   Examples : Salpa, Thalia, Pegea.

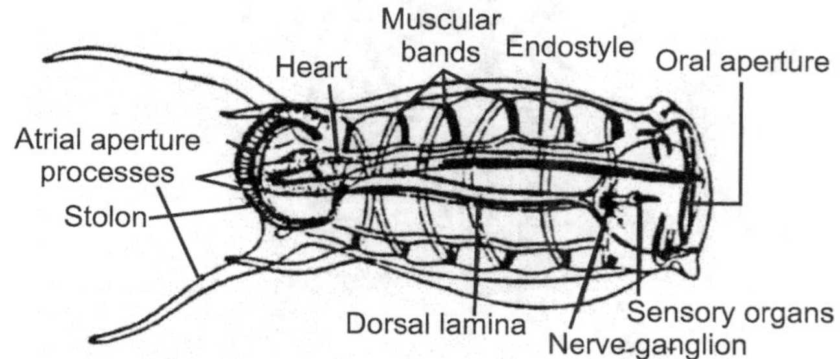

Fig. 1.13 : Structure of Salpa

Class- 3 : Larvacea

1. These are free- swimming, pelagic tunicates.
2. These are neotenic, larval like froms with a persistent notochord.
3. Pharynx possesses only two gill-slits that open directly to outside.
4. Atrium and atriopore are absent.
5. Test or houses is a temporary covering with sieve like apertures and performs the functions of ciliary feeding.
6. Intestine is normally coiled to the right of the oesophagus.
7. No budding occurs.

   This class is divided into two orders: Endostylophora and Polystylophora.

Order (i) Endostylophora

1. Pharynx with endostyle.
2. The house is bilaterically symmetrical with separate inhalant apertures.

   Examples: Appendicularia, Oikopleura.

Order(ii) Polystylophora

1. Pharynx without endostyle.
2. House is biradially symmetrical having a single aperture.

   Example: Kowalevskia

1.4 CEPHALOCHORDATA

- They are also marine and filter-feeders.
- The notochord extends up to the cephalic or head region and persists throughout the life.
- The nerve chord persists throughout the life, but **no brain** is formed.
- Excretion occurs by **solenocytes**.
- The gill slits are numerous and persist in the adults. They open in atrium and true gills are absent.
- The body wall consists of **myotomes**.
- Tail persists throughout the life

Examples: *Amphioxus*

Fig. 1.14 : Amphioxus

EXERCISE

1. Give an account of general features of protochordata.
2. Give the general characters of subphylum urochordata and cephalochordata with suitable examples.
3. Write short notes on:
   (a) Fundamental chordate characters
   (b) Urochordata
   (c) Cephalochordata
   (d) Cyclostomata
   (e) Agnatha
4. Write an essay on origin and ancestry of chordates?
5. Give an account of different theories proposed on origin of chordates.
6. Write short notes on:
   (a) Barrington's hypothesis
   (b) Echinoderm ancestry
   (c) Denterostome line of chordates evolution
   (d) Urochordate ancestry.

☆☆☆

2

CHAPTER

AGNATHA

2.1 INTRODUCTION

The term agnatha is derived from two Greek words *'a'* means not and *gnatha* means jaws. The animals included in agnatha are having circular mouth without jaws. Hence, they are commonly called cyclostomes. Cyclostomes include Lampreys and hagfishes. Both are aquatic forms but lamprey is fresh water form and Hag fishes are exclusively marine forms. They may be parasitic or scanvengerous.

The body of cyclostomes is elongated, slender and almost cylindrical with laterally compressed tail. The body is divisible into three parts, a head, a trunk and a tail. Median fin is present and is supported by cartilagenous fin rays. The skin is without scales, soft slimy and smooth. The skin is with unicellular mucous glands keeping skin slimy.

2.2 GENERAL CHARACTERS

1. Mouth without jaws, circular and suctorial.
2. Paired appendages and exoskeleton is absent.
3. Body is round and elongated.
4. Single median nasal opening present.
5. Presence of large number of gill slits, from 7-14 pairs.
6. Paired fins absent, single median fin present.
7. These are cold blooded animals.
8. Tongue is muscular and is rasping organ.
9. Heart is two chambered without conus.
10. Long kidneys and long archenephric duct.
11. Genital ducts are absent.
12. Pineal apparatus fairly will developed.
13. Skull has a membranous roof.
14. Fertilization is external.
15. Development is direct or indirect.

(2.1)

2.3 CLASSIFICATION

The group Agnatha is further divided or classified into two classes,

1. Class-Ostracodermi-Ancient armoured fishes
2. Class-Cyclostomata-The living agnatha

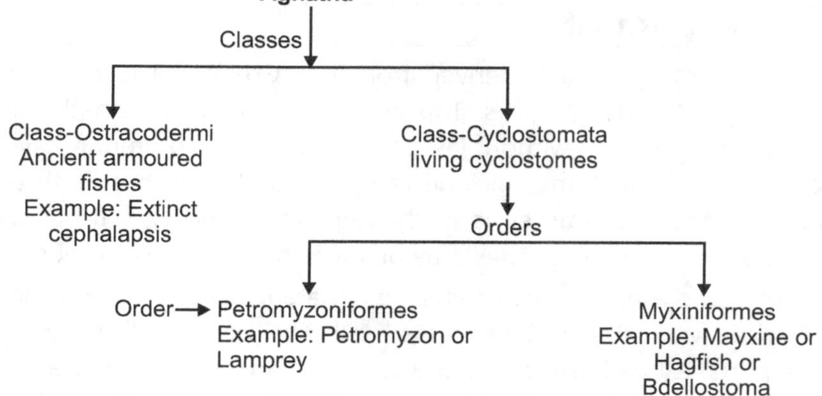

Class-1 : Ostracodermi

Characters :

1. Fossil jawless agnatha of fresh water.
2. They had fish like bodies with bony dermal plate in the skin.
3. Single nostril on the top of the head.
4. Some forms had one pair of fin behind head.
5. Paired eyes with median pineal eye is present.
6. The gill slits are round with gill pouches.
7. Slit like mouth is present at extreme front of the head.

Examples : *Cephalpsis*

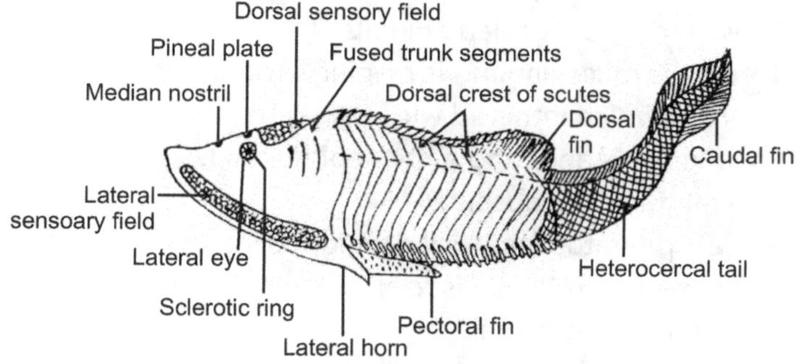

Fig. 2.1 : Cephalpsis

2.4 CLASSIFICATION OF CYCLOSTOMATA

The word cyclostomata has been derived from two Greek words *Cyklos* = Circular and *Stoma* = Mouth. These are the vertebrates without jaws. The mouth is bounded by circular covering. These are only living vertebrates belongs to group Agnatha. The cyclostomes include Lampreys, Hagfishes and slime eels. They represents most primitive vertebrates. There are about 50 species of cyclostomes which belongs to orders Petromyzontiformes and Myxiniformes. The species belongs to these orders exhibits differences with respect to habits, habitat and morphological characters, etc.

Cyclostomata :

General Characters :

1. Body is long, rounded and eel like.
2. Skin is soft, smooth and without exoskeleton.
3. Mouth is suctorial with jaws.
4. Single nostril present on dorsal side.
5. Paired fins and appendages are present.
6. Endoskeleton is cartilagenous.
7. Numerous gill slits are present.
8. Horney teeth present for attachment to host body.
9. Notochord is persistant.
10. Heart is two chambered.
11. Single gonad without duct.
12. Development is direct or indirect.
13. Larval stage is microphagous filter feeder.

Class cyclostomata is further divided into two orders :

1. Order – Petromyzontiformes
2. Order - Myxiniformes

1. Order - Petromyzontiformes

General characters :

1. They are commonly called "**Lampreys**".

2. They are aquatic forms and live both in fresh and marine water.

3. The body is elongated, cylindrical and eel like.

4. Mouth circular and branchial basket well developed.

5. Respiratory organs are gills or branchial pouches.

6. Median dorsal fins and caudal fins are present.

7. The heart is three-chambered.

8. Eggs are small and fertilization is external.

9. Development is direct.

10. Free swimming larva is called Ammocoete.

11. Migrates to river for spawning.

12. Pineal eye is persent.

Examples : Petromyzon or lamprey.

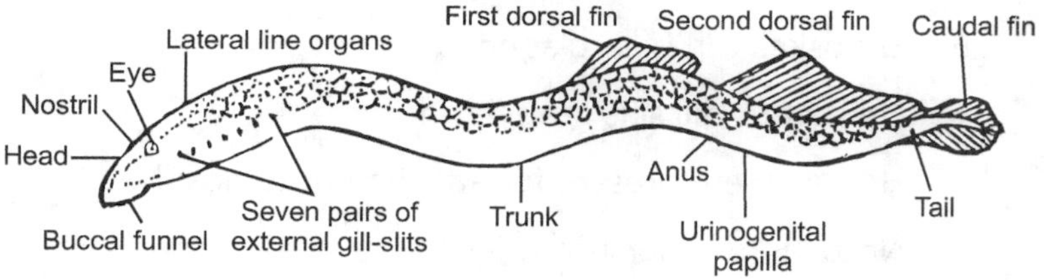

Fig. 2.2 : Petromyzon

2. Order - Mixiniformes

General characters :

1. They are commonly called as **Hag fishes** or **slime eels**.

2. Exclusively marine forms and buried in the sand.

3. Body is slender "eel like" and feeble.

4. Dorsal fin is poorly developed.

5. Three or four pairs of tentacles present around mouth.

6. Tongue well developed with small teeth.

7. Single nostril lying very close to mouth.

8. Six to fifteen pairs of gill pouches open directly into gut.

9. The eggs are large and development is direct.

10. Branchial basket is not well developed.

11. Urinogenital sinus is absent.

12. Pineal eye is absent.

Examples : Myxine or slime eel, Bdellostoma or hag fish.

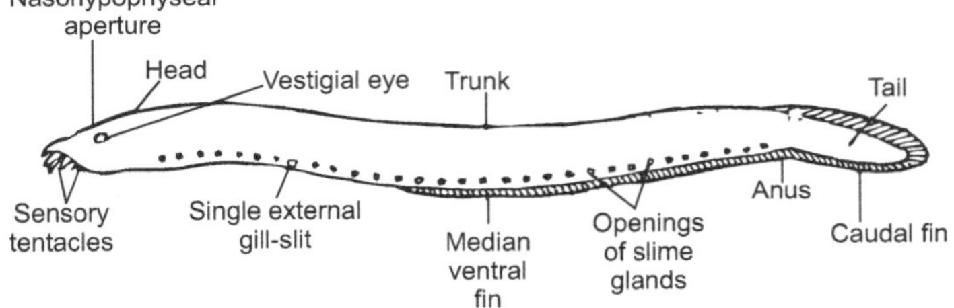

Fig. 2.3 : Myxine

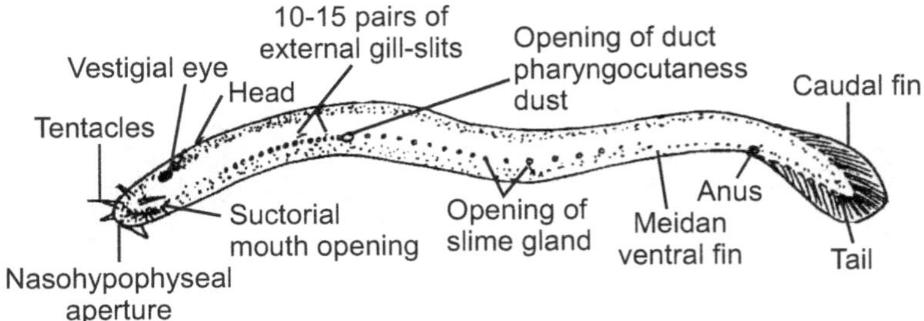

Fig. 2.4 : Bdellostoma

EXERCISE

1. Give an account of general organisation, general characters, habits and habitat and distribution of cyclostomata.

2. Describe the geographical distribution, habits and external features of *Petromyzon*.

3. Write short notes on:

    (a) Agnatha

    (b) Cyclostomes

    (c) Petromyzon

    (d) Myxine

    (e) Adaptations in Petromyzon

    (f) Economic importance of Petromyzon and Myxine

    (g) Habits and habitat of Myxine

4. Give an account of classification of Agnatha.

3

CHAPTER

PISCES

3.1 INTRODUCTION

The name Pisces for the fishes is derived from the Latin. Fishes are the vertebrates evolved in Silurian period of palaeozoic era. Fishes increased their number in Devonian period, hence Devonian period is called as age of fishes. Fishes are the first jawed vertebrate. They are well adapted for aquatic life living in both fresh and marine water.

3.2 GENERAL CHARACTERS

1. Fishes are aquatic animals, living both in fresh and marine water.
2. Body is streamlined or spindle shaped, helpful for swimming.
3. Exoskeleton on body surface is in the form of placoid, cycloid, ctenoid and ganoid scales.
4. Lateral line system is well developed.
5. Mouth has true jaws with teeth.
6. Locomotion by paired and unpaired fins, supported by true dermal fin rays.
7. They are cold blooded vertebrates.
8. Endoskeleton is cartilaginous or bony.
9. Respiration by gills.
10. Heart is two chambered (1 Auricle and 1 ventricle) and venous heart is present.
11. Adult kidneys are mesonephric and excretion ureotelic.
12. Ribs are present.
13. Sexes are separate and some males have a copulatory organs.
14. Fertilization is external or internal, females are oviparous or ovoviviparous.
15. Development is direct and eggs with much yolk.
16. Migration and parental care are observed.

Examples : Shark, Sawfish, Labeo, Macrel

3.3 CLASSIFICATION

Super class-Pisces is further classified into two classes as follows :
1. Class – Chondrichthyes or Elasmobranchii
2. Class – Osterichthyes or Telostomii

1. Class - Chondrichthyes

General Characters :
1. Mostly marine and predaceous fishes.
2. Presence of cartilegenous endoskeleton.
3. Mouth is ventral.
4. Presence of microscopic placoid scales.
5. Respiration by 5 to 7 pairs of gills.
6. Air bladder and operculum are absent.
7. Heart is 2 chambered (1 auricle and 1 ventricle).
8. Intestine has scroll valve.
9. Tail is heterocercal or asymmetrical.
10. They are cold blooded vertebrates.
11. Ten pairs of cranial nerves.
12. Brain with large olfactory lobes and cerebellum.
13. Sexes are separate and gonads paired.
14. Fertilization internal, they are oviparous or ovoviviparous.

Examples : Sharks, sting ray, Hammer headed Shark, Rat fish etc.

Class – chondrichthyes is divided into two subclasses as follows :
(a) Subclass - Selachi
(b) Subclass - Bradiodonti

(a) Subclass : Selachi
1. Mostly marine.
2. Multiple gill slits on either side protected by individual skin flaps.
3. A spiracle behind the each eye.
4. Cloaca present.

The subclass – Selachi is further divided into two orders as follows :

Order I : Pleurotremata
1. Body is streamlined with heterocercal tail.
2. Pectoral fins are moderate and constricted at base.
3. 5-7 pairs of lateral gill slits are present.

Examples : True sharks, Scoliodon, spiny dogfish, hammer headed shark, whale, shark etc.

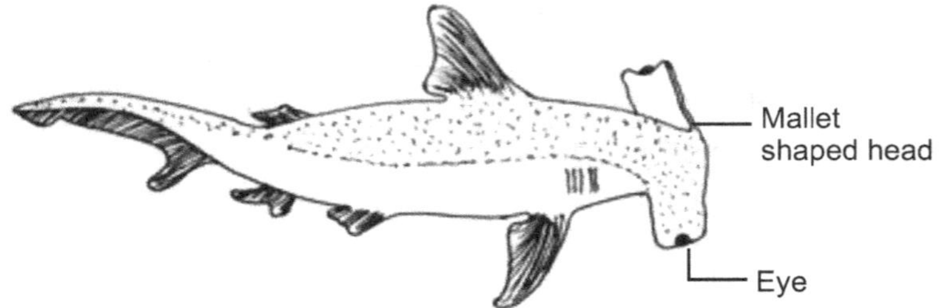

Fig. 3.1 : Sphyrna (Hammer Headed Shark)

Order II : Hypotremata

1. Dorsoventrally flattened body.
2. Gills slits are ventral and five pairs.
3. Pectoral fin enlarged and fused to sides of head and body.
4. Spiracles are large and functional.

Examples : Skates and rays, sting ray, electric ray, eagle ray, guiter fish, Saw fish, etc.

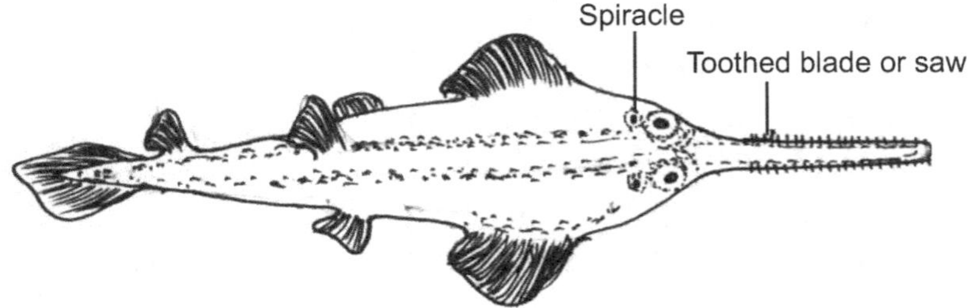

Fig. 3.2 : Pristis (Saw Fish)

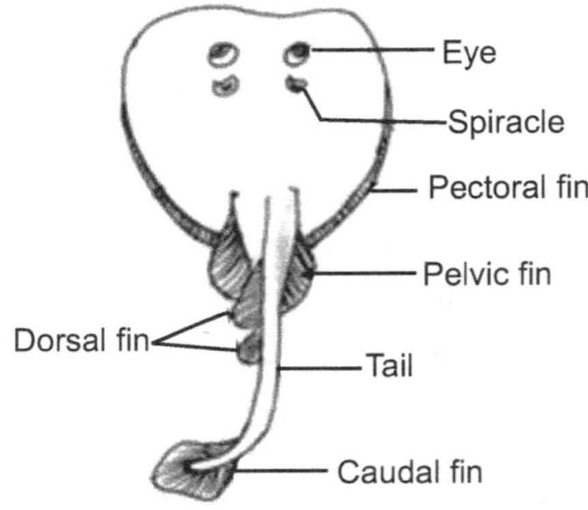

Fig. 3.3 : Torpedo (Electric ray)

(b) Subclass - Bradiodonti

The sub class – Bradiodonti is named as Holocephali also. It include single order named as Holocephali.

Order I : Holocephali

1. Single gill opening covered by operculum.
2. Cloaca absent
3. Operculum present
4. Adults have no scales

    Example : Chimaera (rat fish)

2. Class - Osteichthyes

General Characters :

1. They inhabits in fresh and marine water.
2. Presence of bony endoskeleton.
3. Large cycloid, ctenoid, ganoid scales are present.
4. Mouth is terminal.
5. Four pairs of gills and gill slits covered by operculum.
6. Tail is homocercal and rarely diphycercal.
7. Air bladder is present.
8. They are cold blooded animals.
9. Heart is two chambered.
10. Intestine has no scroll valve.
11. Adult kidneys are mesonephric. Excretion ureotelic.
12. Brain with small olfactory lobe, small cerebrum and well developed optic lobes and cerebellum.
13. Ten pairs of cranial nerves.
14. Mostly oviparous but few are viviparous.
15. Fertilization is external (except few).

    Examples : Rohu, Catla, Eels, Carp, Sea horse, Anabus etc.

The class – Osteichthyes is divided into two subclasses as follows :

(a) Subclass - Sarcopterygii

(b) Subclass - Actinopterygii

(a) Subclass : Sarcopterygii

1. Paired fins are lobed with fleshy bony central axis.
2. Two dorsal fins present.
3. Caudal fin heterocercal with epicordal lobe.
4. Popularly called fleshy or lobe finned fishes.

It is divided into two orders : Crossopterygii and dipnoi.

Order I : Crossopterygii

1. Paired fins lobate
2. Caudal fin 3 lobed.
3. Air bladder vestigeal.
   Examples : Latimeria (Living fossil)

Order II : Dipnoi

1. Air bladder single or double, lung like
2. Median fins continuous to form diphycercal tail
3. Internal nares present
   Examples : Neoceratodus, Protopterus, Lepidosiren

(b) Subclass : Actinopterygii

1. Paired fins are thin and broad or without fleshy basal lobes.
2. One dorsal fin is present.
3. Caudal fin without epicordal lobe.
4. Popularly called as ray finned fishes.

It is divided into three super orders : Chondrostei, Holostei and Teleostei.

Super order A : Chondrostei

1. Tail fin heterocercal.
2. Scales usually ganoid.

Order I : Polypteriformes

1. Rhomboid ganoid scales.
2. Dorsal fin with 8 or more finlets.
   Example : Polypterus

Order II : Acipenseriformes

1. Skeleton is cartiloganous
2. Scaleless
   Examples : Ascipenser, Polyodon

Super order B : Holoseti

1. Ganoid or cycloid scales present
2. Tail fin is heterocercal

Order I : Amiiformes

1. Cycloid scales are overlapping.
2. Long dorsal fin.
   Example : Amia

Order II : Semionotiformes

1. Ganoid scales are in oblique rows.
2. Body elongated.
3. Snout elongated

   Example : Lepidosteus

Super order C : Teleostei

1. Mouth terminal
2. Cycloid scales, ctenoid or absent
3. Tail fin mostly homocercal
4. Swim bladder are present
5. Spiracle is lost

Order I : Clupeiformes

1. Scales cycloid.
2. Fins without spines.
3. Tail fin homocercal.

   Examples : Notopterus, Esox, Clupea

Order II : Scopeliformes

1. Swim bladder absent.
2. Dorsal and anal fin without spines.

   Example : Harpodon

Order III : Cypriniformes

1. Body may be scale less.
2. 3rd and 4th vertebrae do not fuse with each other.

   Examples : Carp, Labeo, Catla, Botia, Clarius, Mystus etc.

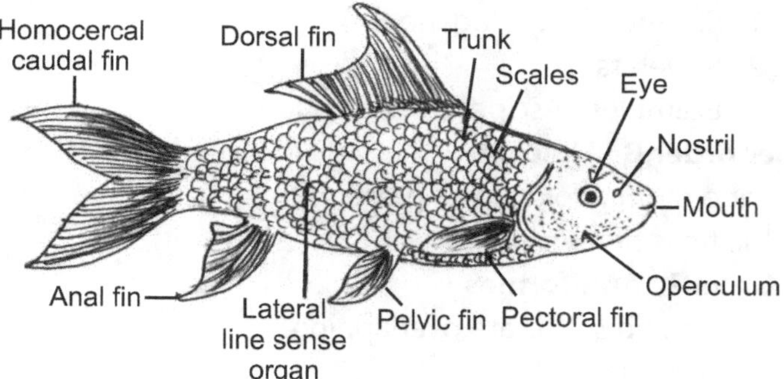

Fig. 3.4 : Labeo

Order IV : Anguiliiformes

1. Body long, slender and snake like.
2. Scales absent or vestigial.
3. Dorsal and anal fins are long.
   Example : Anguilia

Order V : Beloniformes

1. Cycloid scales.
2. Pectoral fin is large and high on body.
   Examples : Exocoetus, Cypselurus, Belone etc.

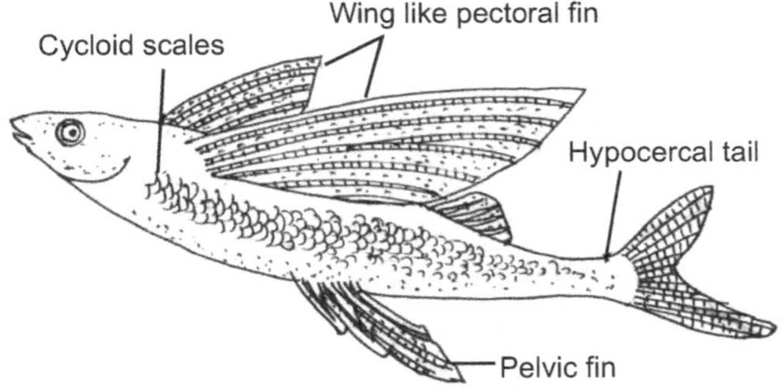

Fig. 3.5 : Exocoetus (Flying Fish)

Order VI : Syngnathiformes

1. Bony rings on the body.
2. Swim bladder is closed.
3. Male has brood pouch.
   Examples : Hippocampus, Syngnathus, Fistularia

Order VII : Ophiocephaliformes

1. Plates like scales on head.
2. Air bladder long and without duct.
   Example : Ophiocephalus

Order VIII : Synbranchiformes

1. Body elongated, eel or snake like.
2. Paired fins, fin rays are lacking
   Examples : Amphipnous, Synbranchus

Order IX : Mastacembeliformes

1. Body is eel like.
2. Dorsal, caudal and anal fins united.
   Examples : Mastacembelus, Macrognathus

Order X : Perciformes

1. Two dorsal fins present.
2. Air bladder without duct.

    Examples : Anabus, Perca, Lates, etc.

Order XI : Scorpaeniformes

1. Well developed head and pectoral fin.
2. Gill covered by spines

    Example : Pterois

Order XII : Pleuronectiformes

1. Body flat
2. Head asymmetrical
3. Swim bladder absent

    Example : Flat fishes

Order XIII : Echeneiformes

1. First dorsal fin has flat adhesive disc.
2. Head possess sucker.
3. Cycloid scales are present.

    Example : Sucker fish

Order IVX : Tetraodontiformes

1. Strong jaws.
2. Scales are spiny.

    Examples : Globe fish, Trunk fish.

Order XV : Lophiiformes

1. Luminescent organs absent.
2. Flexible dorsal spines.

    Examples : Lophius, Antennarius

3.4 RESPIRATION IN FISHES

Fishes are aquatic animals living in both fresh water and marine water habitat. In fishes, respiration is carried out by different structures. The main respiratory organs in fishes are gills. The gills in chondrichthyes (Cartilagenous fishes) and osteichthyeses (Bony fishes) differs with respect to their number and structure. In addition to gills some fishes may contain accessory structures for respiration, these are collectively called accessory respiratory organs in fishes.

The structure of gills in cartilagenous fishes and bony fishes are discussed in detail in this chapter.

(1) Structure of Gills in Cartilaginous fishes :

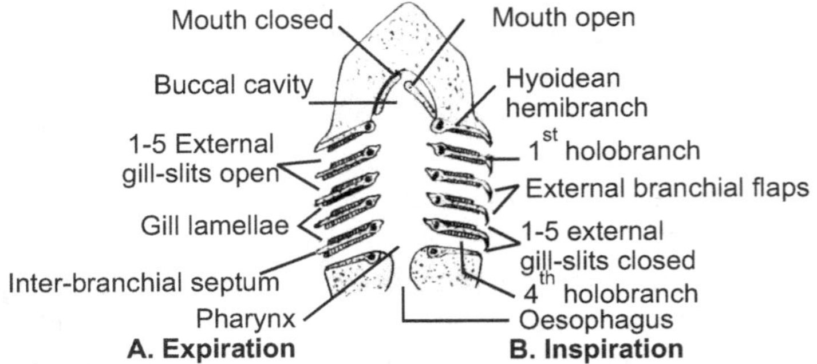

Fig. 3.6 : Scoliodon (respiratory mechanism)

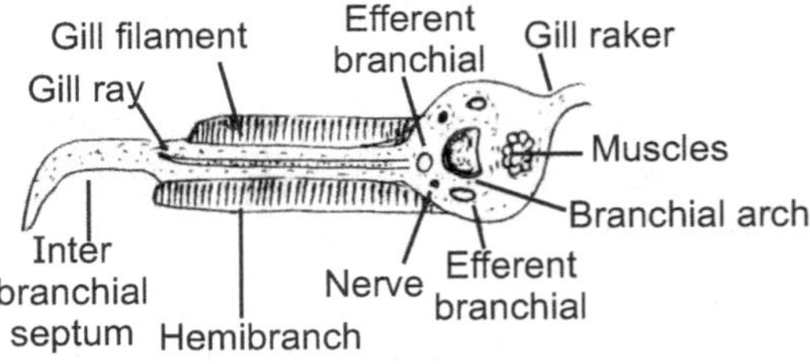

Fig. 3.7 : T.S. of holobranch (scoliodon)

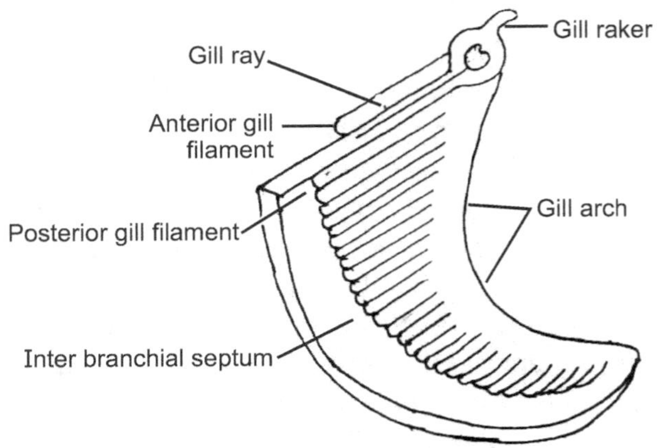

Fig. 3.8 : A part of Holobranch of scoliodon

In cartilagenous fishes, there are five pairs of lateral gill pouches situated in the lateral walls of the pharynx. These are arranged in a series on either side. Each gill pouch is compressed and communicates with pharynx by a large internal brachial aperture. It opens outside by a narrow external brachial aperture (commonly called gill-slits). The adjacent gill pouches are separated from each other by vertical muscular partitions called interbranchial septa or gill septa. The inner part of each septum is supported by visceral arch. From arch cartilagenous gill rays arises in a single row and enters interbranchial septum for further support. Gill racker, a rigid structure is projected towards pharynx and helps to prevent entry of food particles in the inter branchial aperture.

The endodermal mucous membrane of gill pouches raises into a series of horizontal folds to form branchial or gill lamellae. The branchial lamellae are highly vascularized and the exchange of gases takes place between gill lamellae and sea water during respiration. Each gill pouch has two sets of gill lamellae one on anterior wall and other on posterior wall. Each set of lamellae is a half gill or hemibranch hence, each gill pouch has two hemibranches. The gill arch, interbranchial septum and two hemibranch together forms a complete gill or holobranch. The posterior hemibrach is larger than anterior hemibranch.

In cartilagenous fishes, the interbranchial septa extend beyond the gill filaments forming flaps. The flap protect the gill filament as well as the external branchial aperture. The first four branchial arches bears holobranchs and fifth arch has no branchial lamellae. Hence, in scoliodon, there are nine hemibranchs on each side.

(2) Structure of Gill in Bony Fishes :

In bony fishes, there are five pairs of gill clefts which opens by internal branchial apertures in pharynx. The gill clefts are externally covered by a structure called as operculum. The chamber formed by operculum is called a gill chamber on each side. The gills tie inside the gill chamber. It opens posteriorly by a crecentic opening, as external branchial aperture. The operculum is internally lined by a branchial membrane which extends behinds upto body wall. It acts here as a value. It regulates unidirectional flow of water i.e. from gill chamber to the exterior and prevents backflow.

The gill of bony fishes lacks the interbranchial septum which is present in cartilagenous fishes. The endodermal mucous membrane of gill clefts produces gill filaments. The gill filaments of one side forms a hemibranch. Each hemibranch has a supporting gill ray. The two hemibranchs with its branchial arch together form a complete gill or holobranch.

The branchial arch produces comb like gill rackers towards the pharyngeal side. Due to absence of interbranchial septum, the two hemibranchs are free from each other. Hence, a gill appears fork like. Such type of gill is called filiform or pectinate gill. In this type of gill, each hemibranch is completely comes in contact with water. Every gill filament internally bears several transverse plates called lamellae. The lamellae are highly vascular i.e. supplied with blood vessels. The exchange of gases takes at lamellae in gills.

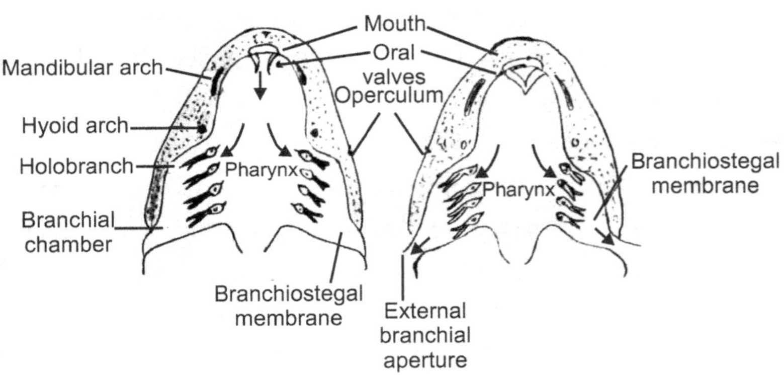

(a) Inhalant current   **(b) Exhalant current**

Fig. 3.9 : Respiratory organs of labeo

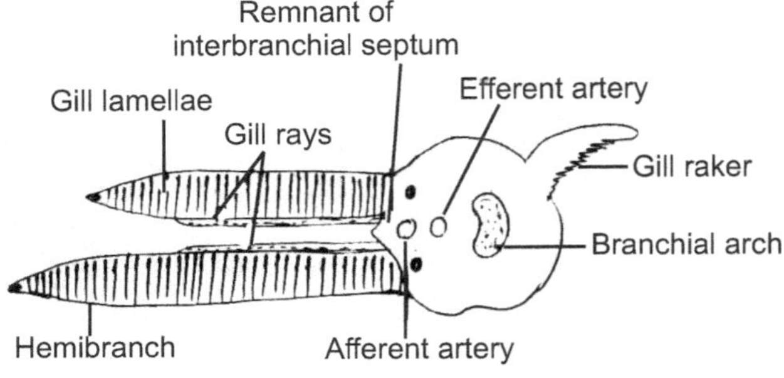

Fig. 3.10 : T.S. of holobranch of labeo

Mechanism of Respiration :

In fishes, respiration is aquatic. The oxygen dissolved in water is utilized for the exchange of gases during respiration. During respiration water from mouth enters the gill pouches by the action of muscles in buccal cavity. The opening of oesophagus is closed and wall of pharynx forces the water in gill pouches through internal branchial apertures. The water passes over branchial lamellae and goes out through external branchial apertures. The branchial lamellae are supplied with blood capillaries hence, exchange of gases takes place here. The oxygen enters the capillary blood by endosmosis and carbon dioxide goes into water by exosmoses. The exchange takes place between water and blood. The oxygen is then transported to all parts of body by circulatory system of fish and water containing carbon dioxide is passed out of body through external branchial apertures. The respiratory movements are caused by pharyngeal muscles which are innervated by cranial nerves.

Accessory respiratory organs in fishes :

Majority of fishes have gills as respiratory organs which are very efficient in gaseous exchange. In some tropical fishes, gills are replaced by specialized respiratory structures called accessory respiratory organs. In some fishes, gills are supplemented by these organs. The specialized structures are developed to overcome the oxygen deficiency in water or to take short excursions on land. In some species of fishes, new structures have developed for accessory respiration, these are called neo-morphic organs. These are found in fishes which are living in deoxygenated water bodies such as stagnant waters, swamps, hill streams etc.

The accessory respiratory organs are :

(1) Skin :

In some fishes, the skin becomes highly vascular and is kept moist by mucous secreted by mucous glands. The exchange of gases takes place between blood of skin and air. e.g. Anguilla, Lepidosiren etc.

(2) Bucco-pharyngeal epithelium :

In most of the fishes, the epithelial lining of buccal cavity is usually highly vascular and permeable to gases in water. Hence, acts as respiratory organ. In fishes like mud skippers the bucco-pharyngeal epithelium is highly vascular and helps in absorbing oxygen directly from the atmosphere.

(3) Gut epithelium :

In the several fishes, the epithelial lining of gut becomes thin and vascular and modified to serve as respiratory organ. It may be present behind stomach, intestine or rectum. e.g. Misgurus, Hoplosternum, Gobitus etc.

(4) Pharyngeal diverticula :

In some fishes, suprabranchial cavities are formed in the roof of the pharynx. These cavities are highly vascularized and serves as respiratory organs. Exchange of gases takes place through these surfaces. e.g. Channa.

(5) Opercular chamber :

In fishes like Periopthalmus and Boleopthalmus the opercular chamber serves as respiratory organ. The opercular cavity in these fishes buldges to form two little balloon like structures which are highly vascular. The gases exchange takes place in balloons.

(6) Branchial diverticula :

The branchial chamber produces outgrowths in some fishes which are complicated than usual pharyngeal outgrowths. These outgrowths or diverticulae serves as respiratory organs. In Indian Catfish, *Heteropneusts fossilis* inhaled air enter the suprabranchial chamber and gases exchange take place there. e.g. Heteropneustes, Clarias, Anabas etc.

(7) Air Bladders :

Air bladder is an outgrowth from the oesophagus in bony fishes. It may retain or lose the communication with the oesophagus. In physostomous fishes, it rises to the water surface and gulp air into air baldder where exchange of gases takes place. Lung fishes stores air in the air bladder for use in respiration during aestivation in burrows.

EXERCISE

1. Give an account of general features of Pisces and classify the fishes upto orders.

2. What is osmoregulation? How freshwater and marine fishes show the osmoregulation.

3. Write short notes on:
    (i)   Cartilaginous fishes
    (ii)   Bony fishes
    (iii)   Dipnoi
    (iv)   Animal diversity in fishes
    (v)   Problems of osmoregulation
    (vi)   Osmoregulators and osmoconfirmers
    (vii)   Osmoregulation in freshwater fishes
    (viii) Osmoregulation in marine fishes.

☆☆☆

4

CHAPTER

AMPHIBIA

The Amphibians were the first vertebrates to invade on land but they occures in fresh water. They are not found in marine water. Amphibians undergoes metamorphosis and generally have two stages in their life, the larval and adult stages. Larval stage is called as tadpole, is fish like, lives in fresh water while adult stage is terrestrial, lives on land. Class amphibia indicates the double life (Gr. Amphi - dual, bios - life).

General characters :

  (1) Amphibians are Aquatic, semiaquatic (fresh water) and terrestrial animals.

  (2) Presence of distinct head, elongated trunk, neck and tail. Tail may be present and absent.

  (3) Presence of two pairs of (tetropad) limbs while some are limbless.

  (4) Paired fins absent. Median fin if present it is without fin rays.

  (5) Skin is soft, moist and glandular. Glands keep skin moist.

  (6) Exoskeleton absent, digits clawless and some with concealed scales.

  (7) Skull is dicondylic i.e. with two occipital condyle.

  (8) Mouth large. Tongue is protrusible. Alimentary canal terminates into cloaca.

  (9) Respiration by lungs, buccopharynx, skin and gills.

  (10) The heart is three chambered having two auricles and one ventricle. The sinus venosus and truncus arteriosus are present.

  (11) Well developed RBCs are biconcave, oval and nucleated.

  (12) Well developed renal portal and hepatic portal system.

(13) Amphibians are the first cold blooded (ectothermic) vertebrates i.e. body temperature is variable (Poikilo thermous).

(14) Kidneys are mesonephric. Urinary bladder is large. Excretion is ureotelic.

(15) Brain poorly developed. Presence of ten pairs of cranial nerves.

(16) Only internal and middle ear and tympanum covered with middle ear.

(17) Lateral line sense organs are present in the larvae and aquatic forms but absent in terrestrial forms.

(18) Sexes separate. Fertilization mostly external. Females mostly oviparous.

(19) Development direct. Cleavage holoblastic.

(20) Ex. Frog, Toad, Salamander, Ichthyophis etc.

Amphibians dominated in the world during Carboniferous period but most of them have became extinct since long. Class – Amphibia is divided into two subclass extinct amphibians are included in subclass – Stegocephalia and all living amphibians are included in subclass – Lissamphibia.

Subclass – Lissamphibia

(1) It includes modern amphibian which lack dermal bony skeleton.

(2) Teeth are small and simple.

It includes three orders. Out of which

Order I : Gymnophiona or Apoda (Gymnos – naked + Ophioneos – serpent like)

(1) They are worm like, limbless and burrowing creatures, commonly called as caecilians.

(2) Tail is short or absent.

(3) Body is smooth, slimy and externally segmented by series of annular grooves.

(4) Dermal scales embedded in skin.

(5) Lidless eyes are reduced and it is covered by skin or maxillary bones.

(6) Limb girdles absent.

(7) Males have protrusible copulatory organs.

Examples : *Ichthyophis; Ureotyphlus, Gegeneophis, Indotyphlus etc.*

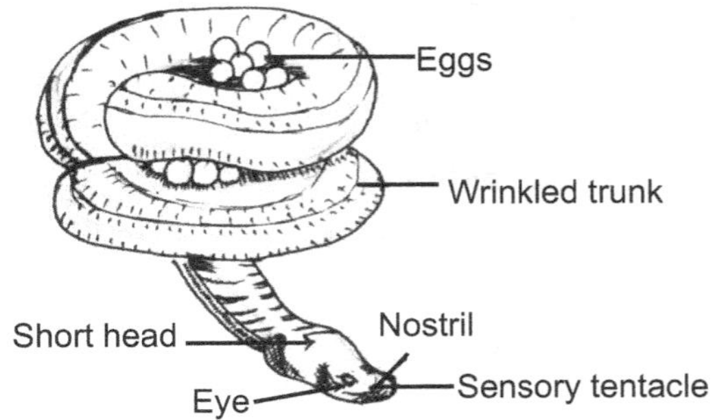

Fig. 4.1 : Ichthyophis

Order II : Urodela or Caudata

(1) Lizard like amphibians with distinct tail.

(2) Skin without scales and tympanum.

(3) Two pairs of almost equal sized limbs.

(4) Gills permanent or lost in adult.

(5)   Larvae aquatic and adult like.

(6) Males without copulatory organ.

(7) Usually oviparous.

Examples : Salamander, Ambystoma, Triton (Newts) etc.

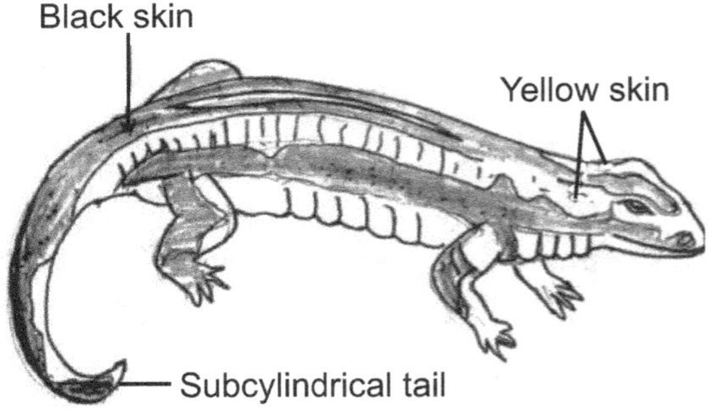

Fig. 4.2 : Salamander

Order – Urodela includes five suborders as follows :

(a) Suborder I : Cryptobranchoidea

Most primitive. Permanently aquatic. Ex. Cryptobranches.

(b) Suborder II : Ambystomatoidea

Adults terrestrial with eyelids. Ex. Ambystoma

(c) Suborder III : Salamandroidea

Three sets of cloacal glands. Ex. Triton and Salamandra.

(d) Suborder IV : Proteidae

Permanently larval form, aquatic and without eyelids.

Ex. Proteus.

(e) Suborder V : Meantes

Permanently larval form and aquatic. Ex. Siren

Order III : Anura or Salientia

(1) Absence of without tail in adults.

(2) Hind limbs usually larger, stout and highly muscular adapted for leaping, jumping and swimming.

(3) Gills are present only in larval stage.

(4) Eye lids well developed and tympanum is present.

(5) Ribs absent, reduced pectoral girdle.

(6) Skin is loosely fitting to body wall and skin without scales.

(7) Fertilization usually external.

Example : Rana, Bufo, Pipa, Hyla, Xenopus, Alytes etc.

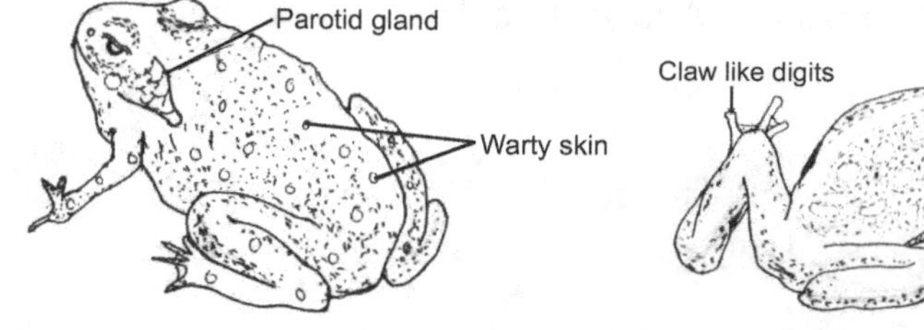

Fig. 4.3 : Bufo   **Fig. 4.4 : Hyla (Tree frog)**

Order – Anura includes five suborders as follows :

(a) Suborder I : Amphicoela

Vertebra amphicoelous. Ex. Leopelma

(b) Suborder II : Opisthocoela

Vetebrae opisthocoelous. Ex. Alytes

(c) Suborder III : Anomocoela

Vertebrae procoelous or amphicoelous. Ex. Pelobates

(d) Suborder IV : Procoela

Vertebrae procoelous. Ex. But-o

(e) Suborder V : Diplasiocoela

First 7 vertebrae procoelous, 8th vertebra amphicoelous.

Ex. Rana, Polypedates

4.2 PARENTAL CARE

Parents take care of the eggs and young ones until they are independent for feeding and take care from enemies (predators) known as parental care. Parental care is very important factor for survival of eggs and young ones.

Animals show great diversity in caring for their eggs and young ones during their development. In general amphibians exhibits more diversity than the other groups. In class – Amphibia, annrans shows much more diversity of parental care than urodales and apodans.

There are two methods involved in parental caring by Amphibia as follows :

(A) Protection by means of nest, nurseries or shelters.

(B) Direct carrying by parents.

(A) Protection by means of nest, nurseries or shelters :

Amphibians have evolved interesting methods to protection of their eggs and young ones from their predators as follows :

(1) Selection of site :

Many amphibians lays eggs in protected places, moist microhabitat on the land. Most of tropical frogs and toads lay their eggs on the land near water. Many tree frogs lay eggs on leaves and branches of trees which are overhanging on the water.

The species of *Phyllomedusa, Rhachophorus Hylodes* etc. lay their eggs on foliage hanging over water. *Rhachophorus malabaricus* and *chiromantis* frogs deposit their spawn on +ve. Frogs select such sites because they are free from aquatic egg predators, the tadpoles on hatching drop down into water bodies to complete their metamorphosis.

(2) Frothing of water :

Some of anurans just after the laying of the eggs, the surrounding water is made frothy by the wriggling movement of hind limbs, so that eggs are prevented from desiccation and also can escape the sight of predators. Both male and female frog participate in this method.

(3) Foam nests :

In the Japanese tree frog, *Rhacophorus schlegeli*, the mating couple digs the hole or tunnel into which eggs are laid in frothy mass to avoid desiccation. During rains, hatching tadpoles are washed down the sloping tunnel into pond or river water for further development.

Female of tree frog *Leptodactylus mystacinus* stirs up frothy mass of mucus and fills it in near water and she lays eggs in them. The tadpoles developing in the nest readily enter in water. While some anurans lays eggs in nests of foam floating on water. When tadpoles hatch they fall into water. In India, foam nest has been recorded in *R. Maculates* by many scientists.

(4) Mud nests :

The tree frog *Hyla fabre*, the male digs a small hole in the mud of shallow water in which female lay her eggs and surplus mud acts as wall of the hole. Tadpoles hatch within thin, relatively safer barrier and develop until they are large enough to fend themselves.

(5) Tree nests :

The South American tree frog, *Phyllomerusa hypochondrales* lay eggs in a folded margin glued together by a cloacal secretion. The tadpoles when formed fall straight into water below.

Another tree frog, *Hyla resinfictrix* female lays eggs in a shallow cavity of tree which is filled with rain water. The eggs hatch and tadpoles develop relatively free from predators.

(6) Shoot nests :

Triton constructs shoot nest on the shoot of trees in which the female lays eggs and the young ones are developed. The whole nest remains covered by a gelatinous secretion.

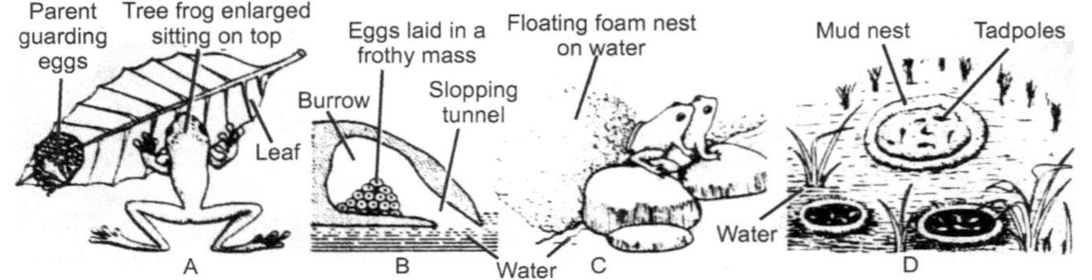

Fig. 4.5 : Parental Care in Amphibia

(7) Direct development :

Some terrestrial or tree frog such as **Hylodes** and **Hyla nebulosa**, the eggs hatch directly into little frogs thus avoiding tadpole mortality and predator attack. In the red blacked salamander **Plethodon cinereus**, the hatchlings are miniatures of the adults.

(8) Defending eggs :

The male of green frog **Rana clamitans** and other species maintain their territories and attack on small sized intruders to defend eggs.

In **mantophryne robusta**, the male actually sits over and holds with hands the elastic gelantinous envelope containing eggs. Some tree frogs laying eggs above water may sit beside the eggs or nest on top of them.

(9) Gelatinous bags :

Phrynixalus biroi large egg gets enclosed in transparent gets gelatinous membranous bag. It is secreted by female.

Salemandrella keyserlingi, a small aquatic salamander also deposits small egg in gelatinous bag which is fasten by aquatic plants.

(B) Direct Carrying by parents :

(1) Coiling around eggs :

Congo eel, Ichthyophis, Gegeneophis, Amphiuma etc. the female lays large eggs in burrows in damp soil. Female guards the eggs by coiling her body until eggs are hatched. Mean while female rotate the eggs.

Salamander, Plethodon also coil round the eggs which are laid in beneath the rock.

(2) Transferring tadpoles to water :

Phyllobates, Artholepits, Pelobates, Dendrobattles, etc. deposit their eggs on the ground. The tadpoles hatching out are transferred quickly on the back of on one of the parents and transported in water.

(3) Eggs glued to body :

Many amphibians carry the eggs glued to their body. **Dusk salamander**, female carries the string of eggs coiled around her neck until eggs hatched. Srilankan tree frog, female lays eggs and male entangles them around the hindlegs. Male carries the eggs until they are hatched. Once hatched the male releases tadpoles into water.

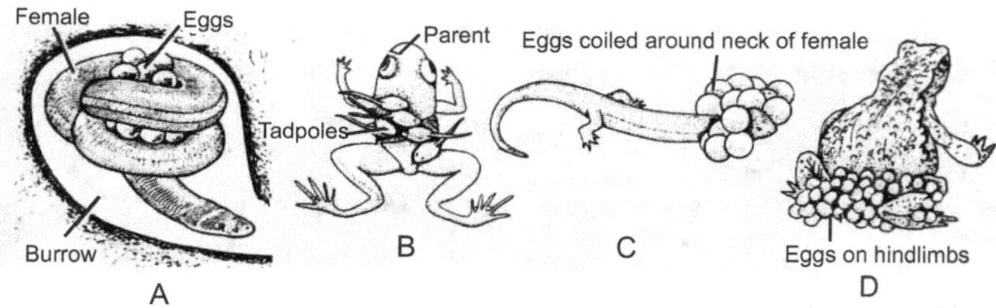

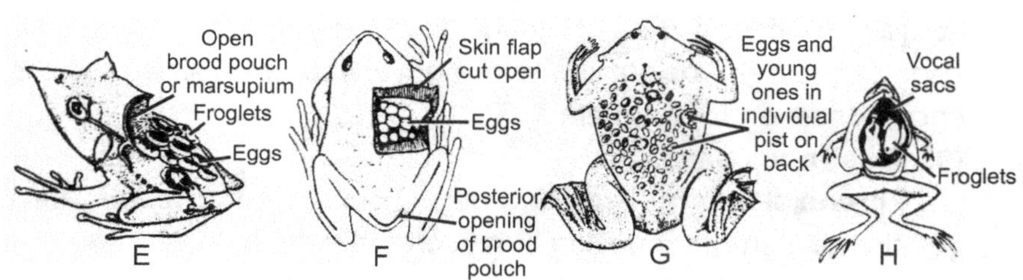

Fig. 4.6 : Direct Parental Care in Amphibia

(4) Egg in back pouches :

The tree frog, **Marsupial** and **toads**, the female carries eggs on her back. The eggs develop into miniature of frogs before they leave the mother's back. **Hyla goeldii**, female lays eggs on posterior part of her back. In **Nototerna**, eggs are covered by skin forming a large pouch which open posteriorly in Cloacal aperture.

(5) Brooding pouches as organ :

In **Rhinoderma darwinii,** It pushes at least two fertile eggs into his large vocal sacs. In vocal sacs complete development take place and emerges out fully formed frog lets. In West African tree frog, the female carries eggs in her buccal cavity.

In **Anthropletis**, male keeps the larvae into his mouth. In Australian frog, the female keeps eggs in her stomach. The tadpoles are expelled to mouth after metamorphosis.

(6) Viviparity :

Some anurans are ovoviviparous. They retain their eggs in the oviduct and female gives birth to living young ones.

African toad, gives birth to little frogs. Europhian salamander produces more small young ones. Vivparity is wide spread in order Gymnophiona. e.g. **Gegeneophis, Geotrypetes, Dermophis, Typhionects**, etc.

EXERCISE

1. Give an account of classification of Amphibia upto orders.
2. Give an account of different types of adaptations found in Amphibia.
3. Write short notes on:
   (a) Class Apoda
   (b) Class Urodela
   (c) Class Anura
4. What is metamorphosis? Give an account of metamorphosis in frog. Add note on significance of metamorphosis.
5. What is parental care? Give an account of parental care in Amphibia.
6. Write short notes on:
   (a) Parental care in Anurans
   (b) Parental care in Urodels
   (c) Parental care in Caecilians.

☆☆☆

5

CHAPTER

REPTILIA

5.1 CLASS - REPTILIA

Reptilia represents the first class of vertebrates adapted for terrestrial mode of life. They are originated from carboniferous amphibian stock and, survived for three hundred million years, dominating on the earth during mesozoic era (the golden age of reptiles), the study of reptile is called as Herpetology (Gr. Herpeton, reptililes). The class reptilia is distinguished by following characters.

General characters :

(1) They are predominantly terrestrial, creeping or burrowing and mostly carnivorous tetrapods.

(2) Body bilaterally symmetrical and divisible into head, neck, trunk and tail.

(3) Skin is dry, rough, cornified and devoid of glands.

(4) Exoskeleton of horny epidermal scales, shields, plates and scutes.

(5) Limbs are short and pentadactyle (except in Snakes).

(6) Clawed digits are present.

(7) Locomotion is by way of creeping or crawling and hence called reptiles.

(8) Mouth is terminal, Jaws, bear simple conical teeth, while in turtles, teeth replaced horny beak.

(9) Alimentary canal terminates into a cloacal aperture.

(10) Endoskeleton is bony. Skull with one occipital condyle.

(11) Incompletely four chambered heart is present (except in Crocodiles).

(12) They are cold blooded (Pikilothermic) vertebrates.

(5.1)

(13) Respiration by lungs throughout life.

(14) Kidney is metanepheric. Excretion is uricotelic.

(15) Amnion, chorion, yolk sac and allantois are extraembryonic membranes around the embryo. Hence reptiles belong to aminiotes.

(16) Male usually with muscular copulatory organ.

(17) Sexes separate, fertilization is internal and mostly oviparous.

(18) Twelve pairs of cranial nerves are present.

(19) Example : Lizard, Snake, Crocodiles, Turtles etc.

Class – Reptila is further classified into four subclasses, according to presence or absence of temporal fossae in the skull.

(a) Subclass I : Anapsida

    (1) Temporal fossa is absent.

    (2) Primitive reptiles with solid skull roof.

(b) Subclass II : Parapsida (Extinct)

    (1) Skull with single lower temporal fossa.

(c) Subclass III : Diapsida

    (1) Skull with two lower temporal fossae.

(d) Subclass IV : Synapsida (Extinct)

    (1) Skull with single lower temporal fossa.

Subclass I : Anapsida

(a) Order : Chelonia or Testudinata

    (1) Body short, broad and oval.

    (2) Limbs clawed, webbed and paddle like.

    (3) Body covered with shell, dorsal carapace and ventral plastron.

    (4) Teeth are absent.

    (5) Cloacal aperture a longitudinal slit.

    (6) Copulatory organ is simple and single.

    (7) About 400 species of marine turtles, fresh water terrapins and terrestrial tortoises.

       Ex. Turtle (Chelone), Tortoise (Testudo)

Subclass II : Parapsida

(1) Single upper temporal fossa present behind the eyes.

(2) All are extinct.

Subclass III : Diapsida

(1) Two temporal fossae are present in the skull.

(2) It includes seven orders, out of which three orders are living.

(a) Order – Rhynchocephalia

(1) Body small. Elongated and lizard like.

(2) Limbs pentadactyle and clawed.

(3) Vertebrae are amphicoelous having intercentra.

(4) Ribs single headed with uncinate process.

(5) Teeth are acrodont.

(6) No copulatory organ in male.

Ex. Sphenodon

(b) Order – Squatmata

(1) It includes present day reptiles.

(2) Exoskeleton of horny epidermal scales, shields and spines.

(3) Skull diapsid. Quadrate movable.

(4) Vertebrae are procolelous.

(5) Temporal fossae may be present or lost.

(6) Quadratojugal is absent.

(7) Squamosal is small.

(8) Teeth accordant or pleurodont.

(9) Male with eversible double copulatory organs.

Order Squatmata is further divided into two suborders.

Suborder – Lacertilia

Suborder – Ophidia

(1) Lacertilia :

(a) Body elongated, lizard like.

(b) Limbs and girdles usually developed.

Ex. Phyrnosoma (Horny toad), Hemidactylus (Wall lizard), Draco (flying lizard), Uromastix, Iguana, Chameleon etc.

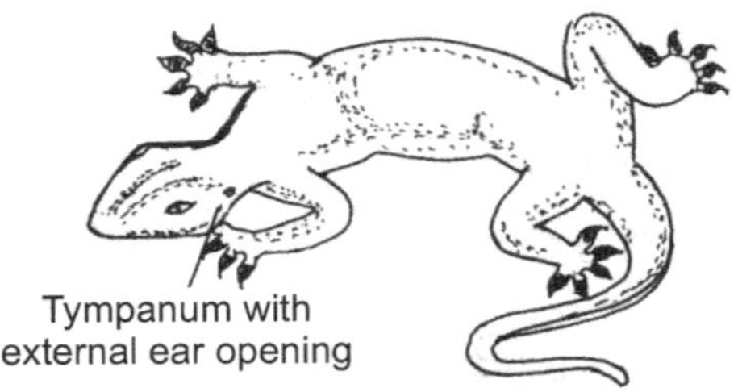

Fig. 5.1 : Hemidactylus (Wall lizard)

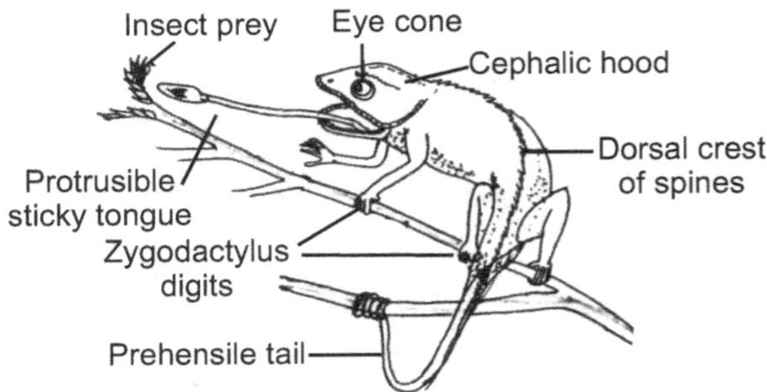

Fig. 5.2 : Chamaeleon

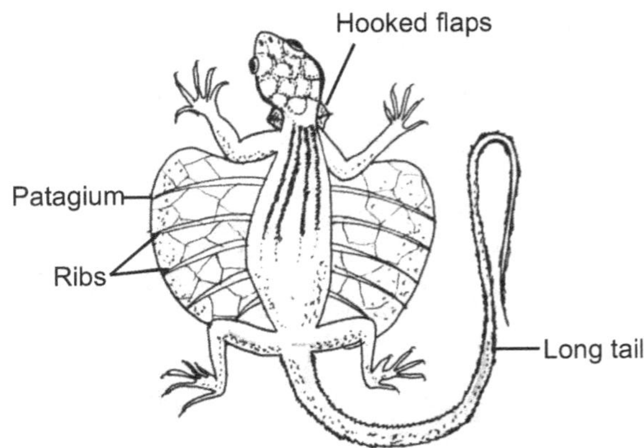

Fig. 5.3 : Draco (Flying lizard)

(2) Ophidia :

(a) Body slender, narrow, snake like.

(b) Limbs absent

(c) Tongue slender, bifid and extensible.

Ex. All snakes – Cobra, Krait, Viper, Python, Dhaman, Typhlops etc.

(c) Order – Crocodilia

(1) Large sized, carnivorous and aquatic reptiles.

(2) Body covered with scales, bony plates and scutes.

(3) Tail long, strong and laterally compressed.

(4) Limbs short but powerful, clawed and webbed.

Ex. Alligator, Crocodiles, Gavialis

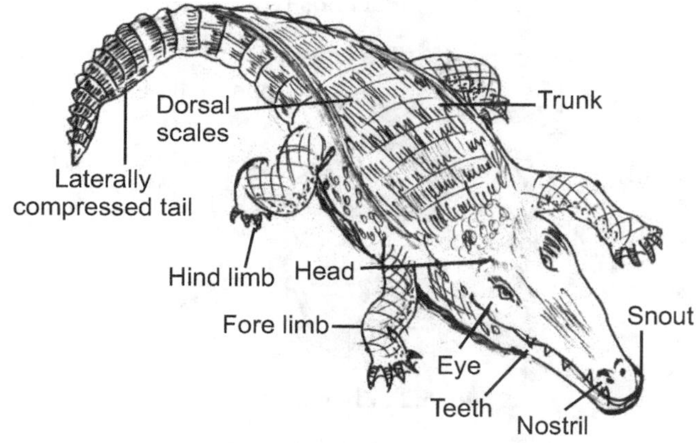

Fig. 5.4 : Crocodylus porosus (Crocodile)

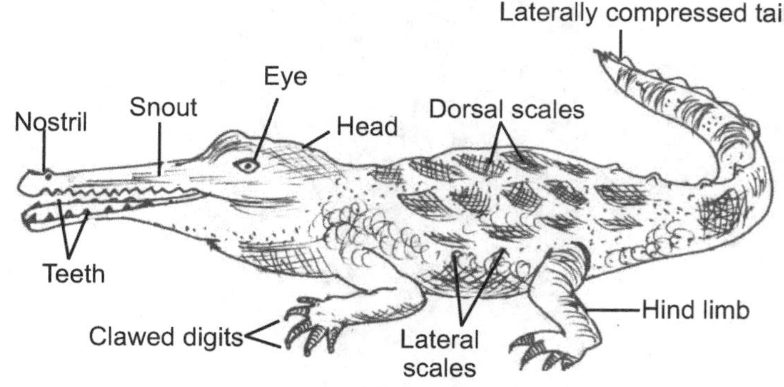

Fig. 5.5 : Gavialis

Subclass IV : Synapsida

(1) This group includes mammal like reptiles (All extinct).

(2) Single lower temporal fossa is present.

(3) A member like cynognathus had two occipital condyles. It was probably hairy and milk producing.

(4) Dentition heterodont type.

Ex. Dimetrodon, Cynognathus.

5.2 VENOMOUS AND NON-VENOMUS SNAKES

The snake belongs to class – Reptilia and subclass – ophidia amongst vertebrates. They are supposed to have evolved from burrowing lizards. Snakes have elongated bodies, without limbs and body covered with scales. They are burrowing in habits.

A majority of snakes are nonvenomous and in fact they are beneficial to the man. Snakes kill large number of destructive rodents and harmful insects but some of the snakes are venomous. In India, certain number of deaths have been recorded from snake bites, even where no venom is injected with the bite. According to W.H.O. (World Health Organization) about 7,000 to 12,000 people dies of snake bite every year in India. Some deaths are due to sheer, fright and some due to ignorance and unscientific method of treatment. Thus the knowledge of various types of snakes and the identification between venomous and non venomous snakes is essential to everybody reducing mortality duet to snake bite. With the help of certain characteristics the venomous snakes can be differentiated from non venomous snakes. It is not an easy task especially for a lay man to distinguish them.

(A) Identification Characters of Venomous snakes from the Non-Poisonous

The snakes are identified by the following characters.

1. The nature of the snake bite mark
2. The nature of flow of blood
3. The nature of the tail
4. The nature of the ventrals
5. The nature of the head shields
6. The nature of the vertebrals.

(1) The nature of the snake bite mark :

If a non-venomous snake bites, there occur many puctures or bite marks of palatine teeth and pterygoid teeth along with the bite marks of maxillary teeth. Thus, two pairs of puncture marks are seen, just like a scratch on the skin of the victim.

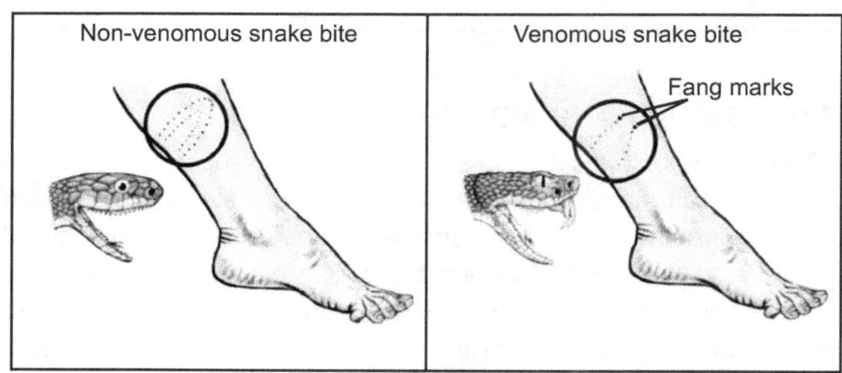

Fig. 5.6 : Biting marks of non-venomous and venomous snakes

When a venomous snake bites then occur many punctures or bite marks of palatine teeth and pterygoid teeth along with fang punctures. The maxillary teeth-punctures are absent, so that only one pair of puncture mark with fang marks seen on the skin of the victim.

(2)   The nature of flow of blood :

When a venomous snake bites then a continuous flow of blood occurs from the wound. The venom prevents the clotting of blood.

If a non-venomous snake bites, the flow of blood is stopped after sometime due to the natural clotting of victim's blood.

(3) The nature of the tail :

If the tail of a snake is flat and laterally compressed, it is a marine, highly venomous snake.

In a terrestrial snake the tail is rounded or cylindrical and not compressed. Snake may be venomous or nonvenomous.

(4)   The nature of the Ventrals :

When the tail is cylindrical or rounded, examine the ventral scales of the snake, i.e. the scales on the ventral side of the trunk.

If all the ventral scales are small or somewhat broad, then it is a non-venomous snake. If the ventral scales are large transverse plates extending completely across the ventral side of belly, the snake may be venomous or non-venomous. In case of some non-venomous snakes, the ventrals are fairly broad but do not extend completely across the belly. E.g. Python.

(5) The nature of the head shields :

If the ventrals are broad, the head shields are examined. If the dorsal scales of the head are small and the head is without shields then it is a viper and is very venomous. If there is a loeral pit present between the eye and the nostril, then it is highly venomous pit viper. If the sub-caudals are double and there is a loeral pit, then it is a russel's viper.

If dorsal side of the head is covered with small scales and large shields, the snake may be venomous or non-venomous. For the confirmation observe the side of the head. If the third supralabial shield touches the nostril and eye, then it is a venomous snake. It may be either cobra, king cobra or coral snake. If the snake has small scales and large shields on the head but there is not loreal pit, and the third supralabial shield does not touch the eye and the nostril, then it is non-venomous snake.

(6) The nature of the vertebrals :

It means by studying the back of the snake and ventral side of the lower jaw. When the middle row of scales on the back called vertebrals are larger than others and the fourth infra-labial shield on the ventral side of lower jaw is larger than the others, then it is a krait. The sub-caudals are single in krait.

If the vertebrals are small and third supra-labial touches the nasal shield and the eye, then it is a venomous snake e.g. cobra.

When the snake has cylindrical tail and elongated broad ventrals, head with shields, small and elongated vertebrals, peculiar markings on the belly, then it is a coral snake.

Venom apparatus : The presence of venomous apparatus is characteristics of venomous snakes. All venomous snakes have a venom apparatus in their head. The venom apparatus is made up of a pair of venom glands, a pair of venom duct and pair of fangs.

(1) Venom glands : The venom glands are modified salivary glands or parotid gland. They are situated one on either side of upper jaw, situated below and behind the eyes. The venom glands are two in number, sac like and almond shaped. It is held in position by ligaments. An anterior ligament attaches the anterior end of the gland to maxilla while posterior ligament extends between the glands and the quadrate. The fan shaped ligaments are situated between the side walls and squamosogudrate junction.

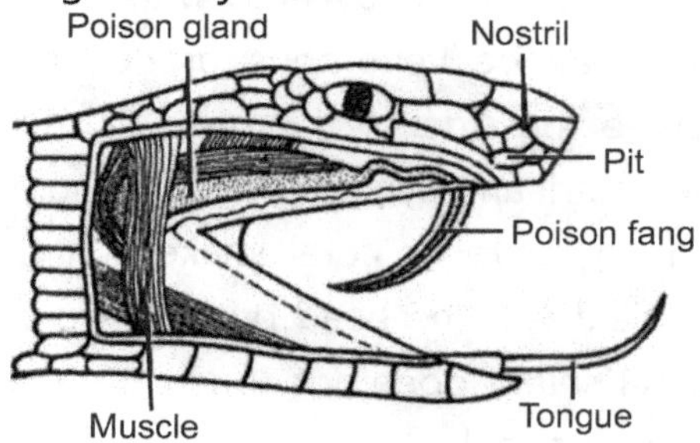

Fig. 5.7 : Poison apparatus of snake

(2) Venom ducts : The anterior end of each venom duct arises into narrow venom duct. Each duct passes forward along the side of the upper jaw and loops over itself just in front of fang. The duct opens either at the base of the fang or at the base of the tunnel on the fang.

(3) Fangs : Venomous snakes have fangs in place of ordinary teeth in front of upper jaw. They are two in number and more or less tubular. They are long, curved, shaped and pointed.

Each fang is like the needle of a hypodermic syringe and thus perfectly adapted for injection of the venom under the skin of the victim. There is one fang on each side but when a fang is lost. It is replaced by a reserve fang, there are two or more reserve fangs enclosed in a sheath. While in nonvenomous snakes there are a number of small teeth's at the place of fang. The fang has a canal or groove they are classified into three types. They are as follows :

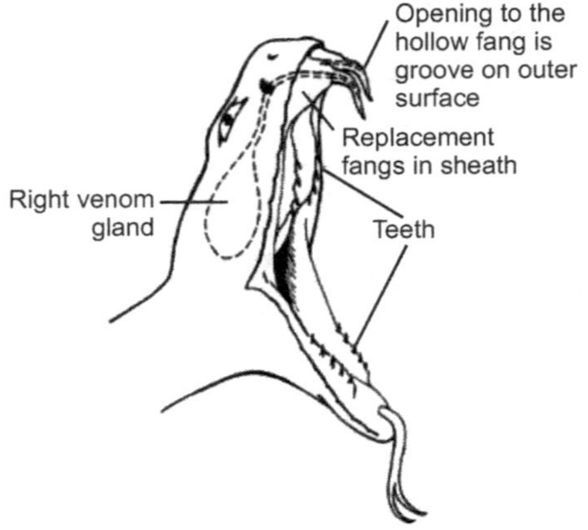

Fig. 5.8 : Fangs of venomous snake

(a) Solenoglyphous : It is hollow fang, it contains a canal extending from base to the tip lined by enamel. Such type of fang is movable when it is not in use, it can be turned inside to lie close to the roof of mouth. E.g. Vipers, Rattle snakes.

(b) Proteroglyphous : Proteroglyphous fang is solid. It has groove from base to the tip along the anterior surface. The fang is permanently erect e.g. Cobras, Kraits, Coral snakes, Sea snakes etc.

(c) Ophisthoglyphous : Ophisthoglyphous fangs are small and solid. It lies at the of maxilla. They contain a groove from base to the tip along back the posterior surface. E.g. Colubrid snakes.

Snake Venom : Snake venom is secreted by certain venom glands present in the head of snakes and injected in the body of victim through the fangs. Venom is clear, sticky liquid of faint yellow or greenish colour. It is tasteless, odourless and acidic in reaction. The pH of Cobra venom is 6.6.4 Russel's viper is 5.8. It contains enzymes, toxins and few metals. The enzymes are phosphates, proteases, hyalurindase etc while metals are Zinc, sulphur and copper etc.

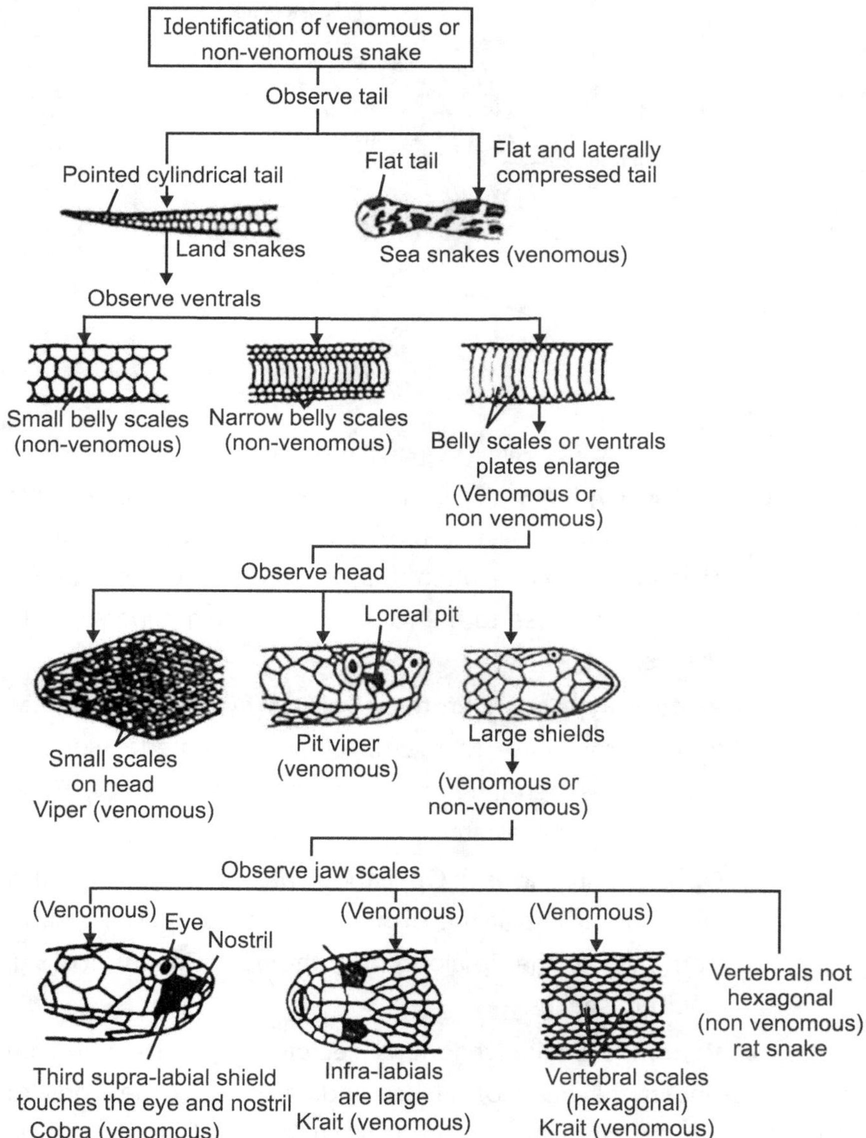

Fig. 5.9 : Identification of venomous and Non venomous snakes

Venom is fatal only when mixed in the blood. The dried venom is thermostable. When dissolved in salt solution or water or glycerin it is fully powerful and poisonous. The venom is a good digestive juice and is capable to digest everything expect hairs and claws. It acts on broken surface of skin or mucous membrane. Thus it is clear that it is toxic only when mixed in blood. Most of the snake's venom consists of both neurotoxins and haemotoxins. Venom of the vipers are haemotoxic which damage the capillaries of endothelium and produces hemorrhages in various tissues ultimately affecting the vascular system. While venom of cobra and Kraits are neurotoxic and acts upon nervous system. The amount of venom discharged by one bite of venomous snakes may vary from species to species. Cobra venom is very powerful.

Antivenom production :

Antivenom production takes place at Haffine Institute, Parel (Bombay). The venomous snakes such as Cobra, Krait, Russel's viper, Saw scaled viper etc are confined. The snake is made to bite on rubber piece which is tied around a small glass vessel. As the snake bites on the rubber, the fangs puncture the rubber. The venom drops collected into the glass a dried in a vacuum desiccator. The dried venom is then dissolved and applied to the horses for immunisation.

The antivenom is the only remedy for a real snake bite. In India, antivenom production is done with the help of immunized horses. A horse is immunized by injecting doses of venom of various snakes into it's body. The doses of venom injected into a horse are low at first and then the dosage is increased gradually over a period of six months or above. If the horse does not show any ill effects, even at a higher dose of venom. It indicates that enough antibodies are formed in his blood. These antibodies help to contract on venom of snakes. The blood from veins of such immunized horses is drawn out and used for cure. The antibodies are present in serum of blood of horses such serum is called antivenom. These antivenoms are finally packed into vials and is specifically used against venom of Cobra, Krait, Russel's viper, saw scaled viper, etc.

Effects of Venom :

The venom of different snakes have different characteristics. The bitten victim may die or recover depending upon the amount of venom injected and its virulence. When the snake bites, if the nervous system of the victim is affected, then the venom is neurotoxic e.g. venom of Cobra, Krait, Sea snake etc. while if the venom affects the circulatory system, then the venom is haemotoxic e.g. Vipers. The effects of venom or symptoms of snake bite in common venomous snakes of India are as follows :

(1) Cobra bite : Venom of cobra is neurotoxic and most virulent. It attacks on nerve centres and causes paralysis of muscles, especially those of respiratory muscles.

Symptoms of cobra bite to the human being show piercing pain, burning sensation, bitten part which turns bluish or blackish, the person suffer from giddiness, weakness in the legs, high pulse rate, speechlessness, dropping of saliva from mouth, closing of eyelids, contraction of pupils, vomiting (nausea) and laboured breathing. Death of the human being within few hours due to failure of respiration or heart.

(2) Krait bite : In case of krait bite, venom directly acts on the brain and spinal cord. It inject large quantity of venom (three times more than Cobra's). Symptoms are very similar to those of Cobra bite except that the unbearable pains in the abdomen due to internal haemorrhage.

(3) Viper bite : Venom of viper is mainly a haemotoxic. It affects on circulatory and nervous system. Symptoms of viper bite includes local pains and swelling; discolouration of bitten parts with acute burning pain. The red fluid oozes out from wound due to massive tissue destruction (necrosis), person feels sickness, vomiting occurs, pupils dialates, person loses consciousness. Death of victim may occur after some hours or even after week due to paralysis of vaso-motar centres and exhaustion from profuse bleeding.

Biting Mechanism :

The skull and jaw bones of venomous snakes are very flexible. They are loosely or movably articulated thus allowing a considerable degree of adjustment during swallowing or striking. In cobras, the fangs are permanently erect but in vipers the large fangs lies against the roof of mouth when closed. Therefore mechanism for biting serves on two main purposes.

(1) erection of fangs

(2) injection of venom into the victim's body

When the viper strikes a victim, the mechanism takes place in a series of movements which occur in a chain. Firstly contraction of digastric muscles, lowers the mandibles so that mouth is open and lower end of quadrate trusts forward. Due to this the pterygoid is pushed forward. It pushes the ectopterygoid upwards. This results, the maxilla bearing fangs to rotate through 90° at the hinge-joint with lacrimal. As a result; the fangs become erect and in the most effective position to strike. A simultaneous stretching of constrictor muscles around the venom gland, forces venom through venom duct into the canal or groove of fang to be injected into victim. When mouth is closed by the contraction of temporal muscles, the above movements are reversed. The fangs embedded in the body of victim are drawn back into the mouth. At the same time vertical fangs rotate to become horizontal.

First Aid Treatment :

First aid treatment means emergency care of the victim just after snake bite, before medical or surgical treatment. The first aid treatment given to the victim are as follows :

(1) In snake bite, victim dies only because of shock rather than effects of venom. Due to shock and emotional upset the rate of heart beat increases, which accelerates the spreading of venom. Hence, first aid treatment given to victim encouragement. It should be free from all fears.

(2) Snake should be identified by observing the bite marks on the body of victims or by observing the snakes if it is nearby.

(3) The victim should not be allowed to move, otherwise the venom spreads with greater speeds in the body of victim.

(4) Immediately after the snake bite, first thing is that to stop blood flow towards heart by applying ligature only in neurotoxin snake bite, a few inches above count. It can be done with handkerchief and cord.

(5) Wound should be washed by using solution of potassium permanganate destroy the venom by its oxidation.

(6) The venom should be neutralized by injecting antivenom as early as possible.

Following are venomous and nonvenomous snakes of India.

(A) Venomous snakes :

    (1) Indian Cobra (Nag) **(Naja naja)**

    (2) King Cobra **(Naja bunarus)** (Nag Raj)

    (3) Common Krait **(Bungarus coeruleus)**

    (4) Banded Krait **(Bungarus fasciatus)**

    (5) Russel's viper (Ghonus) **(Vipera rusalli)**

    (6) Saw scaled viper (Phoorsa) **(Echis carinata)**

    (7) Common green pit viper (Bamboo pit viper) **(Lachesis gramineus)**

    (8) Sea snake **(Hydrophis)**

(B) Nonvenomous snakes :

    (1) Common Rat snake (Phaman) **(Zamenis mucosus)**

    (2) Fresh water Indian snake **(Natrix piscator)**

    (3) Dog faced water snake **(Cerberus rhynopus)**

    (4) Indian Python (Ajgar) **(Python molurus)**

    (5) Earth snake (Sand boa) **(Eryx coricus)**

    (6) Common wolf snake **(Lycodon qulicus)**

    (7) Blind snake **(Typhlops)**

(C) Semivenomous snakes :

    (1) Green Whip snake **(Ahaetulla nasustus)**

    (2) Common Cat snake **(Boiga trigonatal)**

    (3) Brown speckled whip snake **(Dryophis pulverulentus)**

    (4) Flying snake **(Chrysopelea ornata)**

EXERCISE

1. Give an account of the general characters of the class Reptilia and classify it upto subclass level giving suitable examples.

2. Write short notes on:

    (a) Spenodon

    (b) Wall lizard

    (c) Draco

    (d) Chameleon

    (e) Phrynosoma

    (f) Python

    (g) Naja

3. What is adaptive radiation ? Give an account of different types of adaptations in reptiles of Mesozoic era.

4. What is extinction ? Describe different factors of total extinction of Dinosaurs.

5. Describe aquatic adaptation of reptiles with suitable examples.

6. Give general characters of Reptilia.

7. Write short notes on:

    (a) Origin and evolution of reptiles

    (b) Evolution of reptiles

    (c) Adaptive radiation

    (d) Aerial and aquatic adaptations

    (e) Terrestrial adaptations

    (f) Dinosaurs

    (g) Mesozoic era

    (h) Extinction of Dinosaurs

8. Give an account of extinct reptiles.

9. Describe poisonous and non-poisonous snakes with suitable examples.

10. Describe biting mechanism of snakes.

☆☆☆

AVES

6.1 CLASS-AVES (BIRDS)

Birds are well known group of vertebrate having the covering of feathers over their body. It is more homogenous group than any other class of vertebretes. It shows series of strongly marked characters such as their pleasant voice, colouration etc. distinguish from any other class. The distinguishing characters of class-Aves as follows :

(1) Body is spindle shaped and divisible into head, neck, trunk and tail. Neck is long and flexible while tail is short and stumpy.

(2) Body is covered by feathers.

(3) Skin is loosly arranged, dry with preen or oil glands at root of tail.

(4) Forelimbs are modified into wings for flight.

(5) Hindlimbs are large and adapted for walking, running, perching, swimming etc.

(6) Jaws are modified into horny beak or bill.

(7) Teeth are absent in the beak.

(8) Birds are the first warm blooded vertebrate (Homoiothermous).

(9) Heart completely four chambered. Only right aortic arch is persistant in adult.

(10) Endoskeleton is in the forms of bones. Bones are light and also contain air cavities. Hence, the bones are called pneumatic.

(11) Respiration by lungs. Lungs are fixed and provided with air sacs.

(12) Only left ovary and left oviduct is retained.

(13) Urinary bladder is absent.

(14) Cloaca is present.

(15) Eyes are with sclerotic plates.

(16) Skull is smooth and monocondylic.

(17) Ribs double headed (bicephalous).

(18) Vertebrae of trunk are fused.

(19) Tail is reduced and tail vertebrae fused to form pygostyle.

(20) Birds are urecotelic. Kidneys metaephoric.

(21) Brain large but smooth, Cerebrum, cerebellum and optic lobes greatly developed.

(22) Sexes separate. Sexual dimorphism is distinct.

(23) Fertilization internal. Eggs are large, megalecithal and covered with calcarious shell.

(24) Parental cave and migration in birds are well marked.

(25) Ex. Parrot, Sparrow, Myna, Koel, Hoopae etc.

The class-Aves mainly divided into two subclasses.

Subclass-I : Archaeornithes

Subclass-II : Neornithes

Subclass – Archaeornithes

1. Primitive lizard like birds.

2. Wings primitive, with little power of flight.

3. Each hand bears three clawed fingers.

4. Vertebrae are amphicoelous.

5. Tail with 18-20 free caudal vertebrae.

This subclass includes a single order.

Order – **Archaeopterygiformes**

1. These are primitive lizard like birds to Jurassic period.

2. Fore limbs shows three fingers.

3. Tail is long, prehensile with more than 13 caudal vertebrae.

4. Pygostyle absent

5. Jaws modified into beak but beak with teeth.

6. Body covered with feathers.

7. Hindlimbs pentadactyle with opposable toes.

8. Archaeopteryx shows connecting link between Reptiles and Aves.

Example : Archaeopteryx.

Subclass – Neornithes

1. True birds are included.
2. Metacarpals fused.
3. Second finger longest in size.
4. Thirteen or less caudal vertebrae.

Subclass : Neornithes divide into four superorders

Super order :   A Odontognathae

B Palaeognathae

C Impennae

D Neognathae

Super order : A – Odontoganthae

1. Flightless birds.
2. Teeth are present in socket.
3. Extinct birds.

E.g. Ichthyornis, Hesperonis.

Super order : B-Palaeognathae

1. Flightless birds.
2. These are running birds.
3. Wings and flight muscles are present.
4. Feathers are without interlocking mechanism.
5. Males are with copulatory organ.
6. Legs are long and muscular adapted for running.
7. Eggs are large.

Super order: Palaeognathae is divided into seven orders as follows.

Order-I : Struthioniformes

1. Legs with two toes and strongly developed.
2. Pubic synphysis is present.

Ex. **Ostrich**

Order-II : Rheiformes

1. Legs with three clawed toes.
2. Ischial symphysis is present in vertral side.

   *Ex. **Rhea americana***

Order-III : Casuariiformes

1. Head with comblike structure.
2. Fore limbs are greatly reduced.

   Ex. Emu and Casuarius.

Order-IV : Apterygiformes

1. Wings are absent or greatly reduced.
2. Feathers are simple and bristle like.
3. Long bill with nostrils near tip.

   Ex. Kiwi

Order-V : Dinornithiformes

1. All are extinct.
2. Legs massive, wings absent and beak short.

   Ex. Moas

Order-VI : Aepyornithiformes

1. All are extinct.
2. Wings are tiny.
3. Legs are powerful and 4 toed.

   Ex. Gaint Elephant (Birds of Africa)

Order-VII : Tinamiformes

1. Small runner birds.
2. Sternum is keeled.

   Ex. Tinamus, Eudromia.

Superorder : Impennae

Order-I : Spenisciformes

1. Modern, aquatic and flightless birds.
2. Wings are paddle like.
3. Feet are webbed.

Ex. Penguins

Superorder : Neognathae or Carinatae

1. Mostly modern, usually small sized and flying birds.
2. Well developed wings and flight muscles.
3. Beak is modified in different birds for their mode of feeding.
4. Head and eyes are large.
5. Oil glands are present.
6. Sternum with well developed keel.
7. Pygostyle is present.
8. Male has no copulatory organ.

The super order-Neognathae includes several orders as follows :

Order-1 : Passeriformes

1. This is largest in all bird orders.
2. Feet are adapted for perching type.
3. Three toes directed forward and one backward.

Examples : All perching birds includes Sparrow, Crow, Myna, Oriole, Tailor bird, Bulbuls, Magpie, etc.

Order-2 : Piciformes

1. These are insectivorous wood borers.
2. Beak strong, long wedge shaped and pointed.
3. Tongue is long and barbed.
4. These are good climbers.

Examples : Wood pecker, Sap-sucker etc.

Order-3 : Colombiformes

1. These are grain and fruit eating birds.
2. Beak is slender.
3. These are good fliers.

Examples : Pigeon, Doves etc.

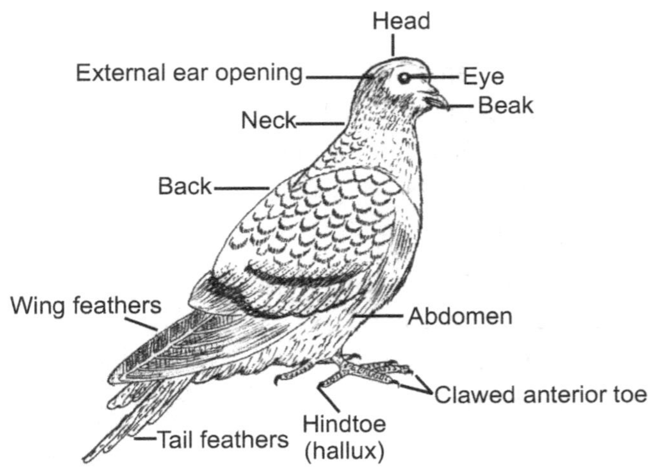

Fig. 6.1 : Pigeon (Columba livia)

Order-4 : Psittaciformes

1. These are fruit eater and colourful bird.

2. Beak is stout, short and curved.

Examples : Parrot, Parakeet

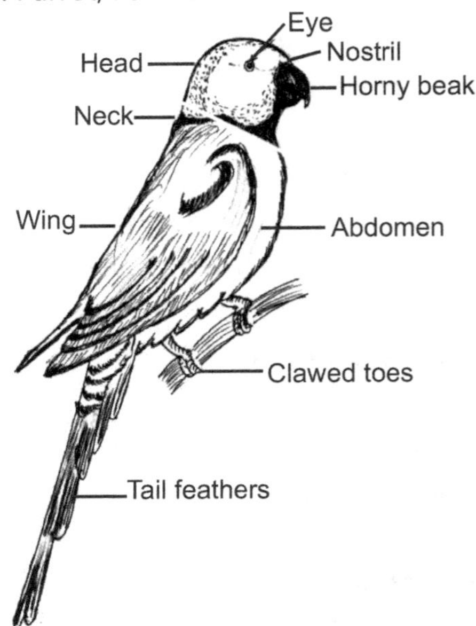

Fig. 6.2 : Parrot

Order-5 : Galliformes

1. It includes gamebirds.
2. Heavy bodied terrestrial birds.
3. Legs strong and good runner.

Examples : Fowl, Peafowl, pheasant etc.

Order-6 : Cuculiformes

1. Wings long and pointed.
2. Beak is slightly curved.

Examples : Koel, Crow-pheasant etc.

Order-7 : Anseriformes

1. Aquatic birds.
2. Beak depressed and tongue thick.
3. Anterior toes are webbed.

Examples : Duck, Swan, Geese etc.

Order-8 : Coraciiformes

1. Beak long, heavy and pointed.
2. Feet are weak.

Examples : Hornbill, Hoopae, Bee eater, Kingfisher, etc.

Order-9 : Colymbiformes

1. They are aquatic.
2. Toes webbed, neck elongated.
3. Beak strong.
4. Found in West and North America.

Examples : Columbus

Order-10 : Preocellariformes

1. Head large and legs short.
2. Good filers.
3. They are large oceanic birds.

Examples : Albatrosses, Petrel etc.

Order-11 : Pelecaniformes

1. Legs short and toes webbed.
2. Beak is modified.

3.   Large aquatic birds.

Examples : Pelican, Little cormorant, Indian darter

Order-12 : Charadriiformes

1.   Legs are long and wading.

2.   Toes are webbed.

3.   Beaks mudprobing.

Examples : Jacana, Sand piper, Gull Curlew etc.

Order-13 : Ciconiiformes

1.   Legs are long and toes are may be webbed.

2.   Beak and neck are very long.

3.   Found in lakes and such places.

Examples : Stork, Heron etc.

Order-14 : Gruiformes

1.   These are living near the shores and banks.

2.   Legs and toes are long.

3.   Poor fliers.

Examples : Crane, Bustard, Common coot etc.

Order-15 : Falconiformes

1.   The diurnal birds of prey.

2.   Beak sharp, strong and curved.

3.   Claws are also strong.

Examples : Flacon, Hawk, Kite, Eagle, Vulture etc.

Order-16 : Strigiformes

1.   These are nocturnal birds.

2.   Head round and large.

3.   Eyes are large and binocular vision.

4.   Powerful feet with clawed.

5.   Beak powerful and pointed.

Examples : Owl, Owlets

Order-17 : Micropodiformes or Apodiformes

1. These are insectivorous birds without legs.
2. Wings long and good fliers.
3. Beak and tail are short.

Examples : Humming birds, swift etc.

Order-18 : Caprimulgiforms

1. Beak weak, short and hooked.
2. These are insectivorous.
3. Long hair like feathers on the sides.

Examples : Goat sucker, night Hawk etc.

Order-19 : Podicipiformes

1. Aquatic birds found in lakes, rivers etc.
2. Hind toes raised.
3. Beak curved.

Examples : Grebe

Order-20 : Phaenicopteriformes

1. Neck very long.
2. Toes webbed.
3. Hallux absent.

Examples : Flemingo.

Order-21 : Opsithocomifores

1. Body long and thin.
2. Wings large but poor flier.
3. Beak strong.

Examples : Opisthocomus hoazin.

Order-22 : Pterocletiformes

1. These are excellent flier.

2.   Beak short and pointed.

Examples : Sandgrouse.

6.2 DIGESTIVE SYSTEM

The digestive system of birds show number of unique features and is adapted for rapid and efficient digestion. The digestive system of birds consist of the alimentary canal and the digestive glands.

(I) Alimentary Canal :

The alimentary canal is long and coiled tube having various diameter. It starts from mouth and end with cloacal aperture. It consists of mouth, buckle cavity, pharynx, oesophagus, crop, stomach, small intestine, large intestine and cloacal.

(1) Mouth : It is bounded by horny upper and lower beaks. Beak is without teeth. It is open in buccal cavity.

(2) Buccal Cavity : Mouth is open in buccal cavity. The floor of buccal cavity is occupied by long of narrow tongue. Tongue is somewhat triangular and pointed at the tip.

(3) Pharynx : Buccal cavity is open posteriorly in pharynx. The paired posterior nares open on its roof. Nares is open in the single median aperture of Eustachian tube and posteriorly. Each tube extends from its opening to the cavity of middle ear. The floor of pharynx has oval aperture i.e. glottis. It leads into trachea.

(4) Oesophagus and Crop : Pharynx leads into long, wide and thick walled tube called as oesophagus or gullet. It runs through neck and open in the stomach.

The middle of oesophagus expands into thin walled, bilobed and elastic sac called as crop. The crop used for reservoir of food. The dry and hard food grains are stored in the crop. The mistened and softened the dry grains takes place in the crop. It also contains the mucus-secreting glands.

Some of birds such as pigeon and some parrots produces milk for nourishment to the young ones.

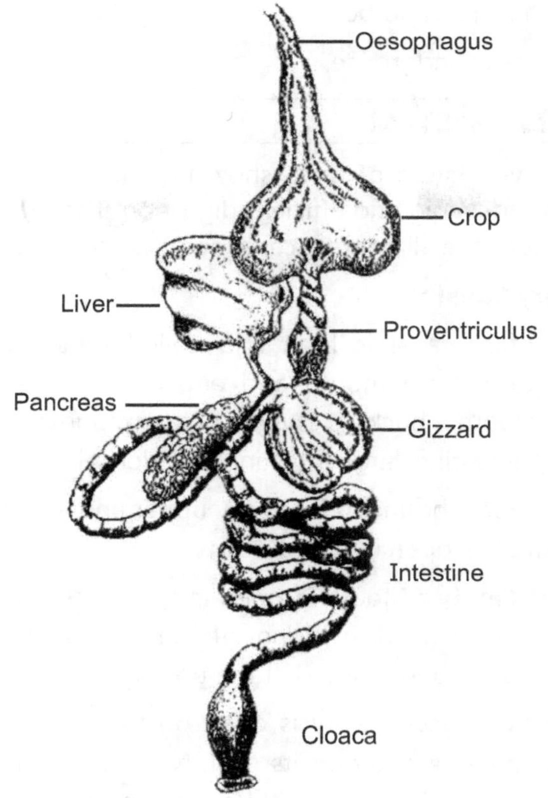

Fig. 6.3 : Digestive system of birds

(5) Stomach : It is divisible into two parts, an anterior proventriculus and a posterior gizzard.

(a) Proventiculus : It is small and thick walled sac. The mucus lining of proventriculus secretes the gastric juice. The spleen is attached to the peritoneum and situated right side of the proventriculus.

(b) Gizzard : Gizzard is the thick walled and muscular. It is laterally compressed and biconvex in shape. Gizzard is thick, rough, horny and green or yellow coloured. The cavity of the gizzard contains grit swallowed by bird. These help the gizzard to grinding the food particles. The gizzard open into small intestine is guarded by pyloric value.

(6) Small Intestine : The small intestine is a narrow tube is about 75 cm long. It is divisible into anterior duodenum and posterior ileum.

    (a) Duodenum : It is distinct U-shaped loop enclosing by pancreas. Duodenum is received by three ducts, one from pancreas and two from the bile ducts of the liver. Internally it is lined by number of villi, crypts of LiberKuhn and goblet cells.

    (b) Ileum : Ileum is very long and extensively convoluted tube have a uniformity in diameter. The inner lining of ileum has numerous, minute finger like projections called as villi. It increases the area of secretion of enzymes and absorption of food.

(7) Large intestine : The small intestine consists of ileum is open into the large intestine. At junction of small and large intestine is pair of small conical blind pouches called as rectal caeca. It is commonly called as hind gut. It is short tube and absorb the some part of water from the food. It is divisible into anterior rectum and posterior cloaca.

    (a) Rectum : It is narrow and about 4 cm in long. It is open in cloaca is guarded by and sphincter.

    (b) Cloaca : The coiled alimentary canal open outside by cloaca. Cloaca is large, prominent and divided into three parts i.e. anterior coprodaum, it receives the rectum. Middle urodaeum, into which opens ureter and gential duct and posterior proctodaum opens outside by the cloacal aperture.

(II) Digestive Glands :

The glands are associated with alimentary canal which secrete different digestive enzymes. It helps in digestion of food material called as digestive glands.

In birds, the digestive glands consists of buccal glands, salivary glands, gastric glands, liver, pancreas, intestinal glands, tubular glands and caecal gland.

(1) Buccla glands : These are present in the buccal cavity. They secrete mainly mucus to moisten the food particles. Probably it secrete enzyme amylase to moisten the food. The tongue also contains mucous glands.

(2) Salivary gland : Pair of angular and unpaired sublingual glands are present in pharyngeal region. Saliva is secreted form these gland. It help in moisten the food particles.

(3) Gastric glands : The epithelial lining of proventriculus shows numerous gastric glands. It secrete the gastric juice containing peptic enzymes.

(4) Liver : It is dark red coloured and bilobed structure. Liver is large, compact and larger right lobe and small left lobe. The two lobes of liver arises two bile ducts, one from each lobe of liver. The left bile duct open into the proximal limb and right bile duct open into distal limb of the duodenum. Liver secretes bile.

(5) Pancreas : Pancreas is compact, reddish gland situated in between two arms of the duodenum. With the help of pancreatic duct it open in the distal limb of the duodenum. Pancreases secretes pancreatic juice contains several enzymes for digestion of food.

(6) Intestinal glands : The lining of small intestine has numerous, microscopic glands called as intestinal glands. They secretes various enzymes.

(7) Tubular glands : These glands are present in internal lining of gizzard. Glands secrete thick, and yellowish green coloured fluid.

(8) Caecal glands : These glands secrete digestive juice for digestion of vegetable fibres and helpful for absorption of water.

(III) Food, feeding and digestion :

Birds are highly selective for their food. They feed on various types of food picked up by beak. The tongue manipulates and lubricates the food grains with the help of the secretion of buccal glands. In birds, teeth absent in the beak, hence swallowed food passes in crop by oesophagus. In the crop food is stored and it is mixed with water, mucus and secretion of buccal glands. So that food grains are softened, and warmth due to body temperature. The wall of the crop does not produce enzymes, then these food grains enter in proventriculus. In proventriculus secretion of gastric glands it help in chemical digestion. In Gizzard, the food grains are chrushed and grind by its muscular contraction, it is more acidic than the crop. It shows little digestion. The partial digestion of food grain here is called as chyme.

Chyme passes into duodenum through pylorus, it is mixed with bile, pancreatic juices and intestinal juices. In birds, the secretion of enzymes and process of digestion is similar to the mammals.

The digested food is absorbed through wall of ileum and the caeca into the body. The undigested part passes into rectum and cloaca. Finally ejected out by cloacal aperture. In birds, faecal matter is semisolid due to reabsorption of the water in the wall of rectum. Birds are uricotelic excrete small quantities of uric acid from facal matter.

6.3 RESPIRATORY SYSTEM

The respiratory system of birds consists of respiratory tracts, respiratory organs or lungs and the air sacs.

(1) Respiratory tract :

It consists of the external naves, nasal sacs, glottis, larynx and syrinx.

- **(a) External naves :** It is commonly called as nostrils. They are pair of oligue slit like aperture situated at the base of upper beak. It leads into short nasal sacs which open directly into the pharynx by the internal naves.

- **(b) Trachea :** It is medium slit like opening called as glottis. Glottis is lies behind the tongue, it leads to wind pipe called as trachea. Trachea antioriorly open by voice less chamber called as larynx. Larynx mostly reduced in birds. Trachea is a long, cylindrical and flexible tube, it is running through neck and in thorasic cavity it expands as syrinx, then it bifurcate into two short bronchi. Both trachea and bronchi supported by series of cartilaginous rings.

- **(c) Syrinx :** It is commonly called as sound producing box in birds. It does not occurs in other vertebrates. Syrinx is situated at posterior end of trachea. Its expanded chamber is called as tympanum. It is supported by last four rings of trachea. Syrinx is characteristics of the birds. The voice is produced by the vibration of tympaniform membranes as the air expelled from the lungs. The pitch of the voice is ultered by changes in the tension of these membranes. While it is absent in some birds like Ostriches, storks and some vultures.

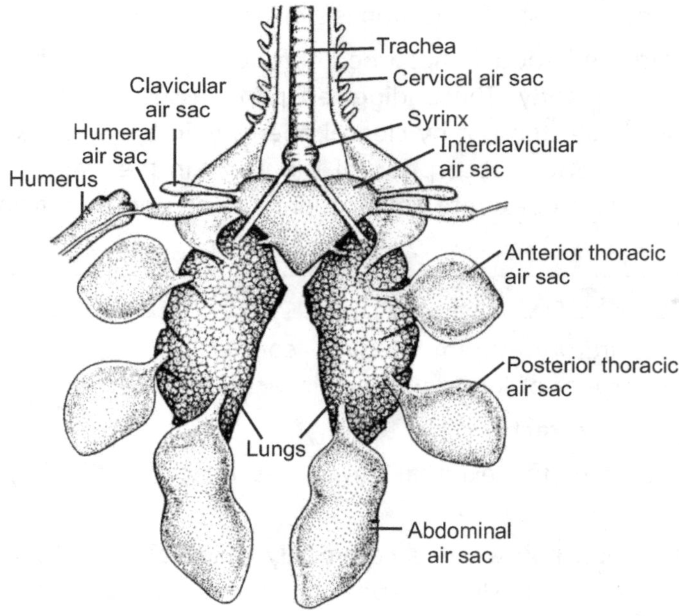

Fig. 6.4 : Respiratory system of birds

(2) Respiratory organs :

It is commonly called as Lungs. The two lungs are small, compact and nonelastic. They are reddish in colour and situated in thoracic cavity but its dorsal surface is closely applied to the thoracic vertebrae and ribs.

The two lungs of birds are solid by spongy structure. It does not store the air. The anterior end of each lung is primary bronchus. The cartilaginous ring reaches upto distal mesobranchus. It sends branches into air sacs and gives off secondary bronchii. It gives again many thin walled tubules forming a loops.

The inner vascular area of air capillaries or tubules are servers as respiratory membranes through which gaseous exchange in takes place. The lungs of birds are more efficient than any other vertebrates for respiration.

(3) Air sacs :

In bird, lungs associated with nine major air sacs as follows :

(a) Inter clavicular : It is a single and somewhat triangular air sac associated with lungs. It gives off an extraclvicular sac communicating with an air cavity of humerus and clavicular sac.

(b) Cervical : A pair of small cervical sacs arise from lung at the base of neck. It cominicate to the neck.

(c) Anterior thoracic : The paired anterior thoracic air sacs present ventral to lungs in the anterior part of chest. They extend upto posterior thoracic air sac.

(d) Posterior thoracic : The paired posterior thoracic air sacs are small in size. They are found posterior part of thoracic cavity. Each posterior air sac is lungs.

(e) Abdominal : the abdominal air sacs are present from outer posterior side of each lung. It is present at dorsal wall of abdomen and coils of the small intestine.

Function of air sacs :

1. They probably acts as balloons to provide the lightness of the body and reducing weight of the body.

2. Air sacs associated with the flight.

3. Air sacs function as accessory respiratory organs. But they serves as reservoirs of air.

4. Air sacs acts as cooling devices in regulation of the body temperature.

5. Probably air sacs function as allow the free movement of heart in the rigid thorax.

6. Air sacs are present two sides of body, it is helpful for maintaining gravity during the flight.

7. Air sacs acts as resonator i.e. force full expulsion of air and control the pitch of sound.

(4) Respiratory Mechanism :

In birds, lungs are relatively small and non distensible, due to rigid skeleton, have large internal surface for gaseous exchange.

In birds, expiration is active process but inspiration is not like other vertebrates. The breathing mechanism like a suction pump.

(a) Ventilation at rest :

When bird take rest, its sternum rises, and falls alternatively by abdominal and intercostals muscles.

During inspiration, the sternum lowered and air sacs expand as a result fresh air enter through nostrils into air sacs. Same time air present in lungs moves to anterior air sacs.

During expiration, the sternum raised and air sacs compressed while lungs expand. Air rushes from posterior air sac into the lungs, then moves air to the outside through nostsrils.

(b) Ventilation at flight :

During flight, sternum is immovable so that air goes in and out of the lungs by elevation and depression of the beak, with the strokes of wings. The migratory birds have more air circulation and gaseous exchange in the lungs.

EXERCISE

1. Give an account of general characters of the class Aves.

2. Give the outline of the classification of the birds.

3. Give an account of the animal diversity in Aves.

4. Describe aquatic and aerial habitat of the birds.

5. Write short notes on:

    (a) Flightless birds

    (b) Flying birds

    (c) Aerial adaptations

6. Define aerial adaptations and describe morphological, anatomical and physiological adaptations.

7. Write short notes on:

   (a) Morphological adaptations

   (b) Anatomical adaptations.

☆☆☆

MAMMALS

7.1 CLASS-MAMMALIA

Mammals are vertebrate animals which constituting the class-Mammalia. They are viviparous i.e. directly gives birth to young ones. The female have well developed mammary glands, which produces milk for feeding to their young ones. The class Mammalia shows large group of animals with diverse habits and habitats. It constitutes highest group of vertebrate animals which highest intellectual capacities. The distinguishing characters of class-Mammalia as follows:

General characters :

(1)   Presence of hairs on the body.

(2)   Presence of external pinna (ear lobe).

(3)   Presence of mammary glands, sweat glands and sebaceous glands.

(4)   Testis are descended into scrotal sacs, situated outside the abdominal cavities.

(5)   Well developed sound producing box e.g. Larynx.

(6)   Internal ear shows highly developed coiled cochlea.

(7)   Presence of well developed cerebrum and cerebellum.

(8)   Presence of corpora quadrigemina and corpus collasum.

(9)   Highly developed parental care.

(10)   Presence of diaphragm in between thoracic and abdominal cavity.

(11)   Alimentary canal terminates by anus.

(12)   Respiration by paired lungs.

(13)   Heart four chambered with double circulation.

(14)   R.B.Cs are non-nucleated, circular and biconcave (In camels and lamas R.B.Cs nucleated).

(15) They are warm blooded animals. (Homoiothermic)

(16) Kidneys are metanephric, excretion is ureotelic and excretory fluid is urine.

(17) Skull dicondylar, skull bones are fused.

(18) Vertebrae acoelous.

(19) Teeth are heterodont, thecodont and monophyodont or diphyodont.

(20) Left aortic arch is persistant.

(21) Sexes separate, sexual dimorphism is distinct and fertilization internal.

(22) Male has erectile and muscular copulatory organ or penis.

(23) Mammals are viviparous except egg laying monotremes.

(24) After birth, young one nourished by milk secreted by mammary glands of mother.

(25) Mammals shows greatest intelligence among other animals.

Class – Mammalia is further divisible into three subclasses

  (a) Subclass I - Prototheria

  (b) Subclass II – Metatheria

  (c) subclass III – Eutheria

(1) Subclass I – Prototheria (Monotremata)

    (a) They are primitive egg laying mammals.

    (b) Mammary glands without teats.

    (c) Testes are abdominal.

    (d) In adult, teeth are absent.

    (e) T shaped interclavicle is present.

    (f) Cloaca is present.

    (g) External pinna (ear lobe) absent.

    (h) Corpus callosum is absent.

    (i) Heart is four chambered.

    (j) Presence of hairy skin.

    (k) They are found in Australia, Tasmania, New Guinea etc.

Subclass – Prototheria has only one living order called – monotremata. The order – monotremata is further divided into two families.

    (A) Family - Ornithorhynchidae

      (1) Upper jaw is produced into beak like resembles with duck.

      (2) Flattened beak.

        Ex. Duck billed platypus

(B) Family – Echidnidae
   (1) Snout is long and narrow.
   (2) Claws powerful and spiny skin.
       Ex. Echidna (Spiny ant eater)

(2) Subclass – Metatheria (Marsupialia) (Pouched mammals)
   (a) Female has abdominal pouch or marsupium.
   (b) Body covered with soft fur.
   (c) Hind limbs larger than fore limbs (Kangaroo).
   (d) Tail long, powerful and muscular.
   (e) Lower jaw is formed of dentary bone.
   (f) Skull is dicondylar.
   (g) Interclavicle bone absent.
   (h) Left aortic arch present.
   (i) Corpus callosum rudimentary.
   (j) Testes descended in scrotal sac.
   (k) Heart four chambered.
   (l) In female, vagina double.
   (m) They are viviparous, immature youngone transfer into marsupium.
   (n) Usually three premolars and four molars in each jaw.
   Ex. Marsupials – Opossum, Kangaroo, Koala etc.

(3) Subclass – Eutheria (Placentalia)
   (a) They are commonly called placental mammals.
   (b) External pinna (ear lobe) present.
   (c) Mammary glands open by the teats.
   (d) Interclavicle is absent.
   (e) Testes extra-abdominal in the scrotal sacs. (except Elephant).
   (f) Corpus callosum is present.
   (g) Separate anal and urinogenital aperture.
   Eutherians are the higher viviparous placental mammals.
   Dentition never exceeds 44. These about 16 living orders.

Order I : Insectivora
(1) They are small, primitive and nocturnal.
(2) Snout long and pointed.
(3) Five toed plantigrade feet present.
(4) Clawed digits present.
(5) They are terrestrial, aquatic, arboreal and burrowing animals.
Ex. Hedgehogs, Moles, Shrews etc.

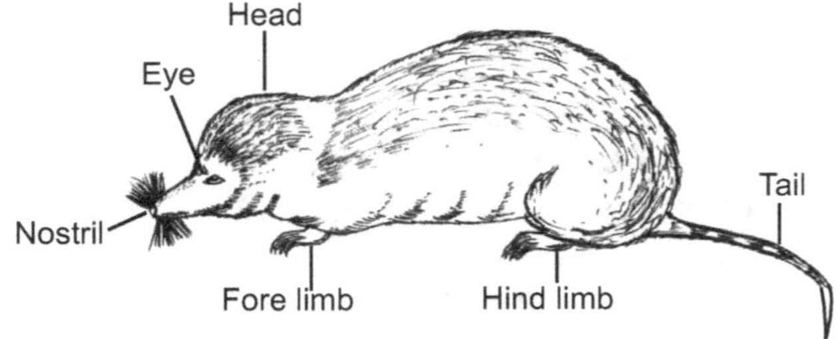

Fig. 7.1 : Shrew (Sorex)

Order II : Chiroptera

(1) These are flying mammals.

(2) Fore limbs modified into patagium.

(3) Hindlegs short and includes wing membrane.

(4) Teeth small, sharp and peg-like.

(5) Eyes small with weak vision.

(6) Ear have large pinna.

(7) Sternum provided with a keel.

(8) They are nocturnal and capable of true flight.

Order – Chiroptera divided into two suborders as follows –

    (a) Suborder - Megachiroptera

        Large, fruit eating bat called flying fox.

        Example – Pteropus, Cynopterus

    (b) Suborder – Microchiroptera

        Small, insectivorous bat

        Example – Megaderma

Fig. 7.2 : Bat

Order III : Dermoptera
(1) Four equal sized limbs.
(2) Commonly called flying lemur.
     Example : Galeopithecus flying lemur.

Order IV : Edentata
(1) Teeth absent or reduced.
(2) Toes with large, strong and curved claws.
(3) Testes are abdominal.
     Ex. Gaint antear, 3 toed sloth

Order V : Pholiodota
(1) Body covered with large overlapping horny scales.
(2) Sparse hairs situated in between.
(3) Tongue long, sticky and protrusible.
     Ex. Pangolin

Order VI : Tubulidentata
(1) Tongue slender and protrusible.
(2) Incissors or canines absent
(3) Skin thick and covered with hairs.
     Ex. Orycteropus

Order VII : Primates
(1) Cerebrum is distinctly developed.
(2) Plantigrade mammal (Locomotion on palm and sole).
(3) Eyes large, directed forward for binocular vision.
(4) Mammary glands with teats present.
(5) Nails present on the tip of digits.
(6) Testes enclosed in scrotum and penis present.
(7) Thumb and great toes are opposable to other digits.
     Exmaple : Lemur, Monkeys, Apes, Man etc.

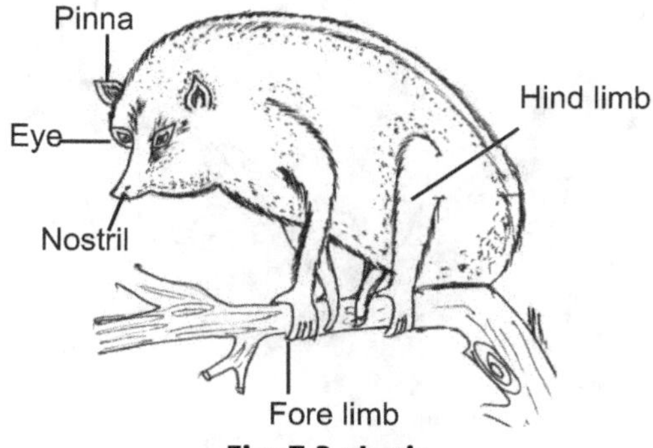

Fig. 7.3 : Loris

Order VIII : Logomorpha

(1) Upper jaw with 2 pairs of incisors, first pair larger than second pair.

(2) Canine absent

Example : Hare, Rabbit

Order IX : Rodentia

(1) Largest mammalian order.

(2) Canines absent

(3) Incissors long and sharp.

(4) Digits are provided with claws.

(5) Jaws elongated forming snout.

Ex. Rat, Mouse, Squirrel etc.

Fig. 7.4 : Squirrel

Order X : Carnivora

(1) Teeth adapted for tearing and cutting the flesh.

(2) Canines large and pointed.

(3) Clawed digits never less than four.

(4) It includes flesh eater mammals.

Ex. Walrus, Seal, Leopard, Cheeta, Hyena etc.

(a) Suborder : Pinnipedia

Marine carnivores

Ex. Seal, Walrus, Sea Lion etc.

(b) Suborder : Fissipedia

Terrestrial carnivores

Ex. Felis tigris, Leopard, Cheeta, Hyena, Mungoose etc.

(c) Suborder : Credontia
   Extint carnivores

Order XI : Proboscida

(1) Snout elongated to form probosic.
(2) Upper incisors are elongated to form ivory tusk.
(3) Skull bone have air cavities.
(4) Hair growth scanty.
(5) Eyes small and ear large.
(6) Testes abdominal.
   Ex. Elephant

Order XII : Hyracoidia

(1) Body small like gmiea pig.
(2) Tail short, ear small and body covered by fur.
(3) Snout is splint.
(4) Herbivorous and native of Africa.
   Ex. Hyrua

Order XIII : Perissodactyla

(1) Cursorial mammal
(2) Stomach simple but Gecum is large size.
(3) Teeth are lophodont having enamel ridge.
   Ex. Horse, Ass, Rhinoceros etc.

Order XIV : Antiodactyla

(1) Even toed mammals.
(2) Upper jaw – incirsors and Canines are lacking.
(3) These are cursorial.
(4) Four chambered compound stomach.
(5) These are cudchewing mammals.
   Ex. Pig., Hippopotamus, Camel, Musk deer etc.

Order XV : Sirenia

(1) Commonly called sea caws.
(2) Body is streamlined, short necked and thick skinned.
(3) Hairs usually absent and pinna absent.
(4) Fore limb modified for swimming pad like.
(5) Hind limbs absent.
(6) Tail posses caudal fin.
(7) They are herbivorous animals.
   Ex. Sea cow, Dungong.

Order XVI : Cretacea

(1) Hairs, claws, teeth, hindlimb, pinnae etc. are either absent or extremely reduced.

(2) Skeleton of forelimbs modified for swimming.

(3) Highly specialized marine mammals.

(a) Suborder : Odontoceti

(1) Homodont and monophyodont teeth present.

(2) Nostril single and olfactory organ absent.

(3) Feed on fishes, shrimps etc.

Ex. Toothed whale, sperm whale, Parpoise etc.

(b) Suborder : Mystacoceti

(1) Teeth absent or rudimentary

(2) Paired nostrils present and well developed olfactory organ.

Ex. Balaena (right whale), Balaenoptera (blue whale)

7.2 CIRCULATORY SYSTEM

Circulatory system of mammals for the distribution digested food, water, gases, hormones, waste materials etc throughout the body. The circulatory system of mammals are closed type. It includes blood, heart, arteries, veins and lymphatic system.

Arterial System :

Arterial system of mammals consists of two main trunks or aortae i.e. pulmonary aorta and systemic aorta; they originate from heart.

(A) Pulmonary aorta : It originates from right ventricle. Which curves upwards and divides into right and left pulmonary aorta. It receives deoxygenated blood and supplies to the paired lungs for purification.

(B) Systemic aorta : It arises from left ventricle. It runs forward and immediately curves to the left. It gives up three main arteries :

(1) Innominate,

(2) Common carotid and,

(3) Left subclavian

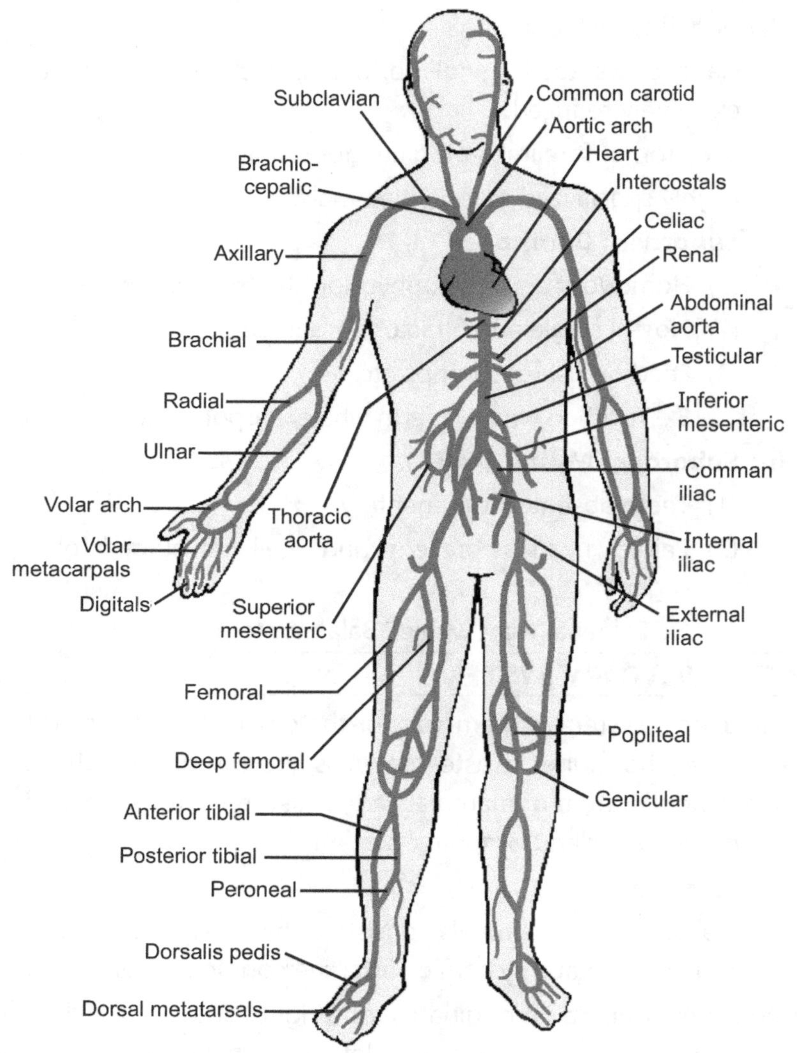

Fig. 7.5 : Arterial system of human being

(1) Innominate : It originates to the right side of arch. It divides into (a) Right common carotid (b) Right subclavian

(a) Right Common Carotid : It originates from innominate. It runs parallel to trachea and passes towards head. It divides into –

(i) Internal Carotid : Which originates from outer side and supply oxygenated blood to the brain.

    (ii) External Carotid : It originates from inner side and supplies blood to the muscles of head, jaw, tongue etc.

(b) Right subclavian : It passes outwards and divides into three branches as follows :

    (i) Right internal mammary : It supplies oxygenated blood to the ventral thoracic wall.

    (ii) Right costo cervical : It supplies blood to the upper part of thoracic wall and neck.

    (iii) Right cervical : It supplies blood to the muscles of neck region.

(2) Left common carotid and Left subclavian : These originates directly from the aortic arch and passes braches similar to the right side.

Dorsal aorta : The systemic aorta curves from left side of heart and forms the median dorsal aorta. In the region of ribs it gives off.

    (i) Intercostals : These are small paired arteries, distributes blood to the thorax.

    (ii) Phrenic : These suppliesblood to the diaphragm.

    Dorsal aorta passes from diaphragm and continues in abdominal cavity, posteriorly it bifurcates into two common iliac arteries. It gives off following arteries.

    (iii) Coeliac : It is median artery and it divides into three arteries as follows.

        (a) Gastric – to the stomach.

        (b) Hepatic – to the liver

        (c) Spleenic – to the spleen

    (iv) Anterior mesenteric artery : It is median and largest, distributing blood to the pancreas, duodenum ileum, caecum and anterior part of colon.

    (v) Renal arteries : These are paired and supplying blood to both the kidneys. It also gives off small branches to the adrenal gland.

    (vi) Gonaidal arteries : These are paired arteries supplying blood, spermatic artery to the testis in the male and ovarian artery to the ovary in the female.

    (vii) Lumbar arteries : These are paired arteries supplying blood to the muscles of the dorsal wall of the abdomen.

    (viii) Posterior mesenteric : It is median ventral artery supplying blood to the colon and rectum.

(ix) Common iliacs : The dorsal aorta posteriorly bifurcates into two common iliacs, each gives vesicular artery to the urinary bladder and ureter, internal iliac to pelvic region, epigastric to the pubic region and external iliac or femoral to the outer region of leg.

(x) Caudal artery : It is small, median artery which is posterior continuation of the dorsal aorta in the tail. It supply oxygenated blood to the muscles of fail region.

Venous System :

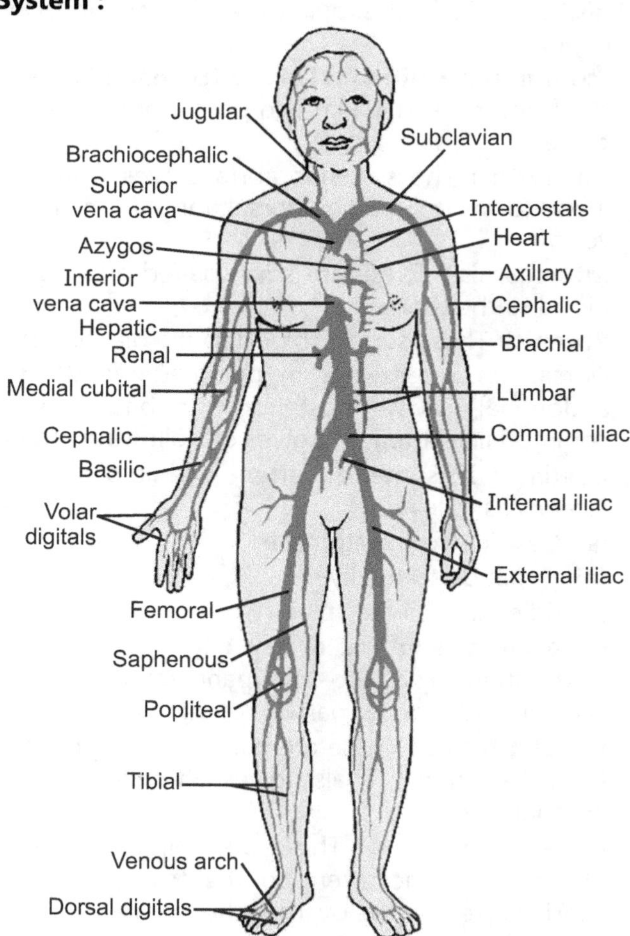

Fig. 7.6 : Venous system of human being

Heart of mammals receives deoxygenated blood by two precavals and one postcaval and enters in the right auricle. While pulmonary vein receives oxygenated blood and enters in the left auricle. The venous system of mammal consists of :

(1) Pulmonary vein : Pulmonary veins collect oxygenated blood from two lungs. They join together forming a common pulmonary vein. It opens dorsally into the left auricle by a common aperture.

(I) R. precavals : Right and left precavals (superior vena cava) collect venous blood from anterior part of the body to heart Precaval is formed by union of the following major branches.

(1) External jugular vein : It is the chief vein collecting deoxygenated blood from head, tongue and jaws by posterior jugular vein and from shoulder region by cephalic vein. Finally joins the external jugular vein.

(2) Internal jugular vein : It collects blood from deeper part of head and brain.

(3) subclavian vein : It runs outward and is formed by union of

(a) Branchial vein : from the heart

(b) Subscapular vein : from shoulder region

(c) Internal mammary : from sternal region

Left precaval (superior vena cava) receives only azygous vein. It is formed by union of intercostal and subcostal veins. They collect deoxygenated blood from lower part of thoracic wall and the phrenic vein from diaphragm.

(II) Postcaval : The postcaval vein or inferior vena cava receives deoxygenated blood to the heart from the posterior region of the body.

(1) Caudal vein : It is slender vein collecting blood from tail and joins the postcaval.

(2) Internal iliac veins : Pair of small veins which collect blood from inner portion of thigh.

(3) External iliac vein : These are pair of large veins, formed by femoral, vesicular and epigastric.

(a) Femoral vein – Outer part of hind limbs

(b) Vesicular vein – From bladder

(c) Epigastric vein – From pelvic region

(4) Iliolumbar veins : A pair of veins from posterior dorsal abdominal wall.

(5) Gonaidal veins : They collect blood from gonads. They are called spermatic veins from testes of male and ovarian veins from ovary of female.

(6) Renal veins : A pair of renal veins collect blood from kidneys.

(7) Hepatic veins : Paired hepatic veins bring blood from the liver, which join the postcaval.

(8) Hepatic portal vein : It collects deoxygenated blood from different parts of alimentary canal and associated organs forming hepatic portal system.

Hepatic Portal System :

Hepatic portal system, collects blood to liver lobes. It is a part of venous system which does not carry the blood directly to the heart. It is portal vein starts from capillaries in one organ and ends into capillaries of another organ. It is made up of the union of several veins as follows :

(a) Posterior mestentric vein : It collects blood from the colon and rectum.

(b) Anterior mesenteric vein : It collects blood from ileum, caecum and anterior part of colon.

(c) Duodenal : It drains blood from duodenum.

(d) Lineogastric : It takes blood from stomach, spleen and pancreas.

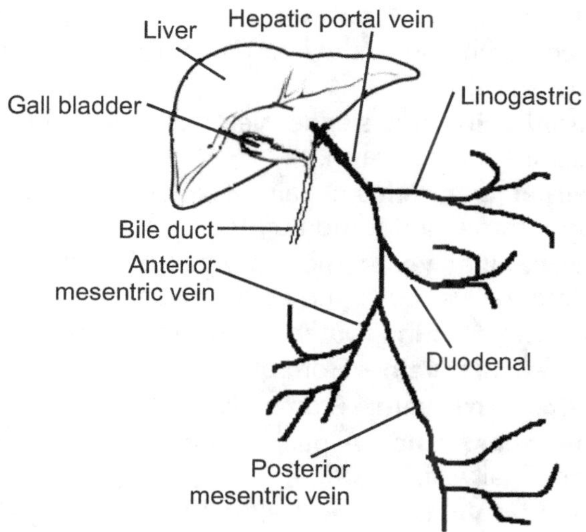

Fig. 7.7 : Hepatic portal system of human being

The hepatic portal vein opens into liver and hepatic vein collects blood from liver and with the help of postcaval vein supplies to the heart.

Lymphatic system :

It is closely associated with blood vascular system. It consists of lymph, lymph sinuses, vessels, nodes (glands) and lymphatic ducts.

(1) Lymph : It is colourless fluid. It's composition is similar to the blood except it has no blood cells. It is derived from the blood and flows through the capillaries in the living tissues, so the name is tissue fluid. It supplies food and oxygen to the cells and tissue.

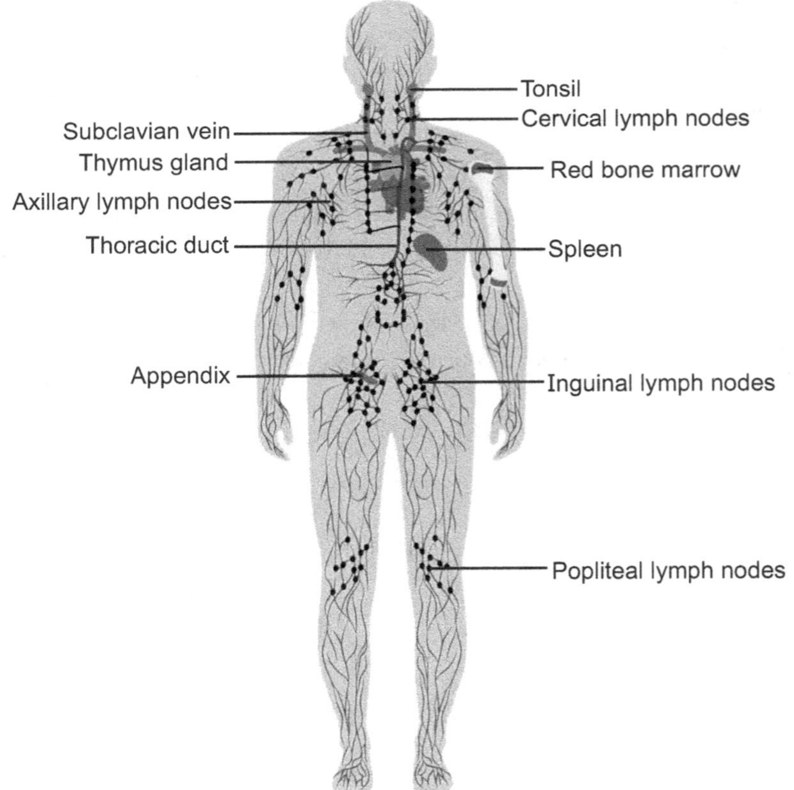

Fig. 7.8 : Lymphatic system of human being

(2) Lymph sinuses : Lymph sinuses are large spaces that occur between the integument and muscles, muscles and bones etc. these are widely distributed throughout the body. Form the sinuses small lymphatic capillaries arise which unite to form the lymph vessels.

(3) Lymph vessels : Lymph vessels are thin walled structures, which possess the valves. Many lymph vessels unite to form lymphatic trunks.

(4) Lymphatic nodes (gland) : The lymph vessels are interrupted at various places by lymph nodes. These are rounded or elongated structures containing lymph vessels, lymphoid tissue, surrounded by dense connective tissue covering. Lymph nodes are named accordingly.

(5) Lymphatic ducts : Lymphatic ducts are formed by union of lymph vessels. These are also called as lymphatic trunks. These trunk empty into the large vein. Thoracic duct is the largest lymphatic duct it lies it lies to left of the spinal cord.

EXERCISE

1. Give an account of the general characters of the class mammals.
2. Given an outline of classification of class mammals.
3. Describe the diversity in mammals with suitable examples.
4. Write short notes on:
   (a) Aerial habitat in mammals
   (b) Terrestrial habitat in mammals
   (c) Aquatic mammals.
5. Give an account of origin of mammals.
6. Write short notes on:
   (a) Origin of mammals through reptiles
   (b) First true mammals
   (c) Polyphyletic origin of mammals.

1

CHAPTER

NUCLEIC ACIDS

1.1 INTRODUCTION

Nucleic acids are the macromolecules containing carbon, hydrogen, oxygen, nitrogen and phosphorous and are present inside the nucleus. The nucleic acid was first isolated by Friedrich Miescher in 1868 from the nuclei of pus cells found on hospital bandages. He called it as **nuclein**. Later, Altman in 1889 gave the term **nucleic acid** to nuclein. It is said to be an acid due to its acidic nature. There are two different types of nucleic acids, namely deoxyribose nucleic acid (DNA) and ribose nucleic acid (RNA). The nucleic acid that contains ribose sugar is called ribose nucleic acid and nucleic acid with deoxyribose sugar is called deoxyribose nucleic acid.

RNA is found in the nucleus as well as in the cytoplasm. In the nucleus, it resides in the nucleolus, nucleoplasm and chromatin. In cytoplasm, RNA forms a part of ribosomes. DNA is found mainly in the nucleus. At resting stage, it is in the chromatin while during cell division it forms a part of the chromosome. DNA is also present in cell organelles like mitochondria and chloroplasts. DNA combines with histone proteins to form nucleoproteins in the nucleus.

1.2 DEOXYRIBOSE NUCLEIC ACID (DNA)

DNA is the most complex molecule found in the cells. It is considered key molecule of cell as it acts as hereditary material in living organisms. DNA is present in all cells except in plant viruses. In eukaryotic cells, DNA is found in the nucleus as a part of chromosomes or chromatin while in prokaryotes and viruses it is single molecule. In bacteria and cell organelles it is circular in form and found free in the cytoplasm. In viruses and bacteriophages they are coiled. DNA in all plants, animals and in many viruses is double stranded and in bacteriophage φ x174, it is single stranded. The number of DNA molecules in eukaryotes corresponds to the number of chromosomes per cell.

1.3 STRUCTURE OF DNA

DNA molecule is an unbranched long chain polynucleotide. It is made up of many monomeric units called nucleotides. Each nucleotide consists of nucleoside and phosphoric acid. The nucleoside is composed of sugar and a base. Each nucleotide is chemically composed of a pentose sugar, a nitrogen base and a phosphate group.

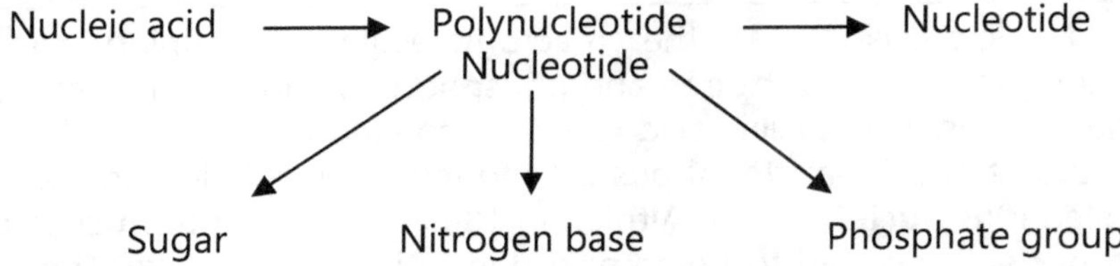

1. Sugar :

The sugar present in the DNA is called deoxyribose sugar. It is a pentose sugar with five carbon atoms (C_5). It contains one oxygen atom less at carbon atom no. 2 (H-C-H) instead of (H-C-OH) in ribose sugar hence called deoxyribose sugar. Out of five, four carbon atoms and one oxygen atom forms five membered ring. The fifth carbon atom is outside the ring and forms a part of a $-CH_2$ group. The sugar molecules along with phosphate group form the backbone of DNA molecule. To the carbon atom number one nitrogen base is attached in DNA.

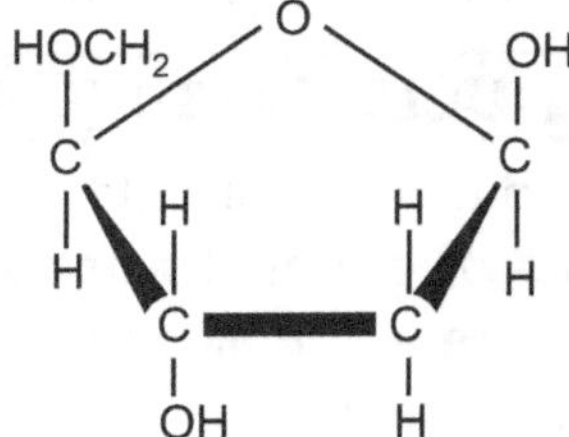

Fig. 1.1 : Deoxyribose sugar

2. Nitrogen bases :

Nitrogen bases are nitrogen containing organic compounds present in DNA. They are of two types, namely purines and pyrimidines. Purines are double ring compounds and pyrimidines are single ring compounds. A purine base is double ring compound with a five membered imidazole ring joined to a pyrimidine ring at position 4' and 5'. The common purine bases in DNA are Adenine (A) and Guanine (G). A pyrimidine base is single ring compound with nitrogen at 1' and 3' position. Pyrimidine base in DNA are Thymine (T) and Cytosine (C).

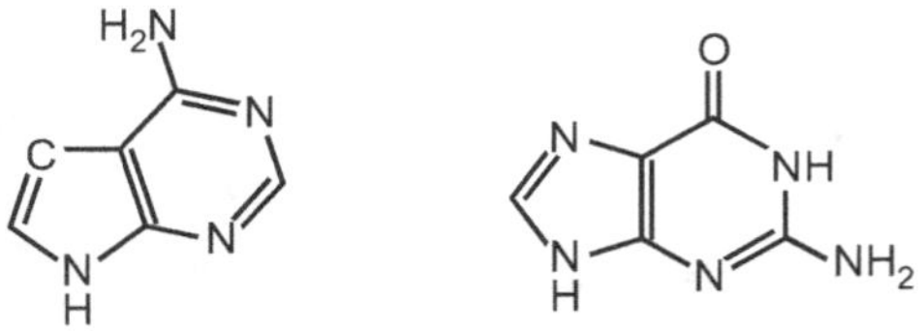

Fig. 1.2 : Purine Nitrogen Bases in DNA

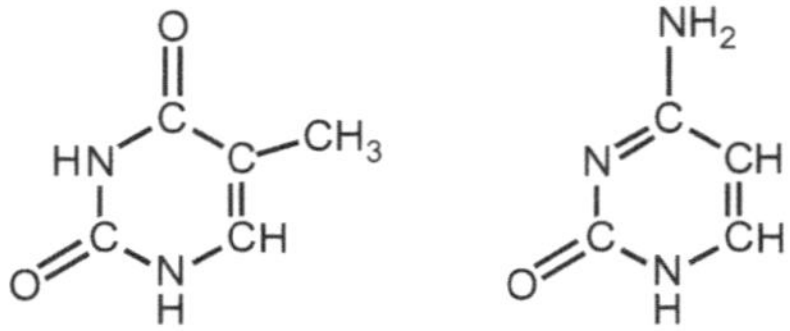

Fig. 1.3 : Pyrimidines Nitrogen Bases in DNA

3.   Phosphoric Acid :

Phosphoric acid (H_3PO_4) is an important component of the DNA. Due to presence of this DNA is acidic in nature. Phosphate group joins with sugar molecules to form phosphodiester bond/linkage. Each phosphate group joins with carbon atom no. 3 of one sugar molecule and carbon atom no. 5 of another sugar molecule to form sugar phosphate backbone of DNA.

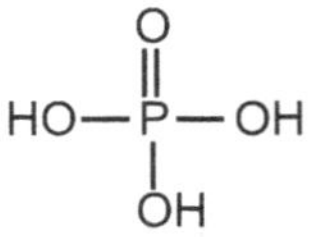

Fig. 1.4 : Phosphoric Acid in DNA

Nucleosides :

Nucleoside is formed by the combination of sugar molecule and a nitrogen base. In DNA deoxyribose sugar combines with any of the nitrogen base from adenine, guanine, thymine or cytosine to form deoxyribonucleoside. In nucleoside, a sugar molecule is linked with nitrogen base through carbon atom no. 1 of sugar with third position of pyrimidines or 9th position of purines. The bond formed between them is called glycosidic bond. Four different types of nucleosides present in DNA are,

1.   Adenosine = Deoxyribose sugar + Adenine

2. Guanosine = Deoxyribose sugar + Guanine
3. Cytidine   = Deoxyribose sugar + Cytosine
4. Thymidine = Deoxyribose sugar + Thymine

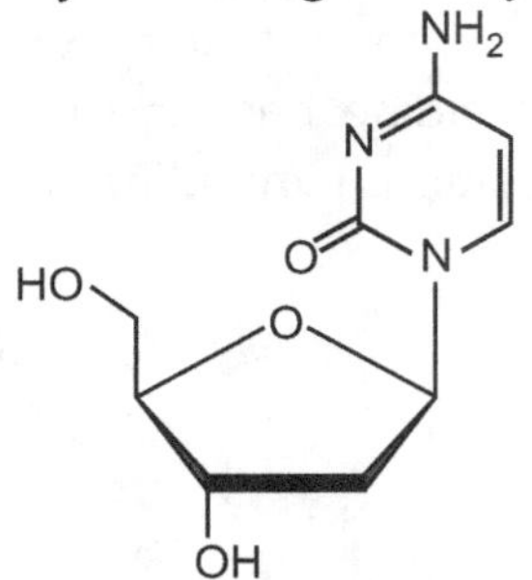

Fig. 1.5 : Cytidine Nucleoside

Nucleotides :

Nucleoside with addition of one phosphoric acid molecule gives rise to nucleotide. The phosphate molecule is attached with sugar molecule at carbon atom no. 3rd or 5th. Thus, nucleotides are all phosphoric acid esters of the nucleotides. These are monomers or basic units of DNA molecule. The nucleotides present in DNA are with deoxyribose sugar hence called **deoxyribonucleotides**. Based on the nitrogen base these are classified into following four types.

1. Nitorgen base + Phosphoric acid = Deoxyribonucleotide
2. Deoxyadenosine + Phosphoric acid = Deoxyadenylic acid
3. Deoxyguanosine + Phosphoric acid = Deoxyguanylic acid
4. Deoxycytidine + Phosphoric acid = Deoxycytidylic acid
5. Deoxythymidine + Phosphoric acid = Deoxythymidylic acid

Fig. 1.6 : Deoxyribonucleotide

1.4 WATSON AND CRICK MODEL OF DNA

The double stranded helical structure of DNA was proposed for the first time by James Watson and Francis Crick in 1953. For this discovery they received Nobel Prize in 1962. The structure of DNA was based on X-ray crystallographic studies carried out by Maurice Wilkins and Rosalind Franklin. The structure has now been widely accepted.

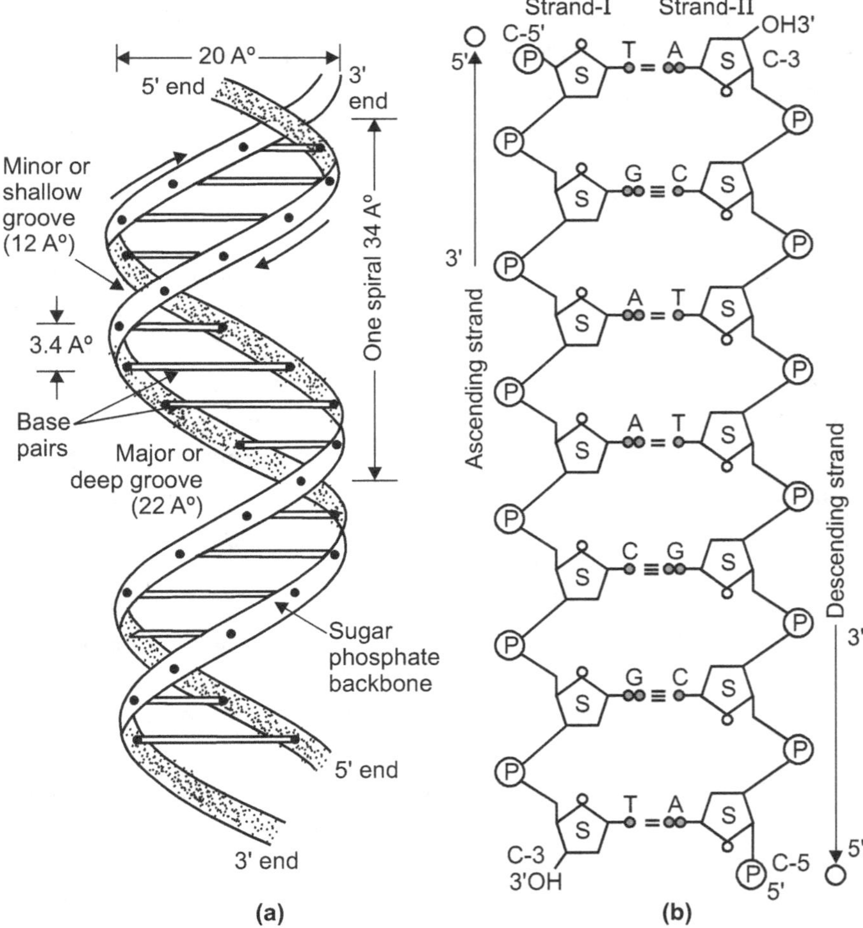

Fig. 1.7 : Structure of DNA (Watson and Crick model)

According to this model, a DNA molecule is linear, double stranded formed of two polynucleotide strands or chains. The backbone of strand is composed of alternate sugar and phosphate arrangement. These are linked by phosphodiester bonds. In successive nucleotides, phosphate group is attached to 5th carbon of a sugar molecule of a nucleotide. The two strands are parallel all along their length and connected together by **rungs** or **transverse steps**. Each strand of DNA molecule is long polynucleotide chain made of units or monomers called **deoxyribonudeotides**. Hence, it is also called polymer compound. Each rung or transverse step is made up of nitrogen bases which are purines and pyrimidines.

The two strands of DNA molecule are spirally twisted (coiled) around each other as well as around common central axis to form a right handed double helix i.e. twisted in a clockwise direction. It looks like a twisted ladder (spiral stair case). The side arms of ladder are called banister or railing and steps are called rungs. The coiling of two strands is **plectonemic** i.e. the two strands can not be easily separated from each other.

In each strand, one end of the strand has one free phosphate group at carbon atom no 5 of the sugar molecule. The end is called as C_5 or 5' (' prime) end. The other end has free –OH group at carbon atom no. 3 of sugar molecule. This end is called as C_3 or 3' (prime) end. Each polynucleotide strand has two ends namely 3' and 5' ends, hence it is called polarized molecule. The strands in DNA are antiparallel i.e. one strand runs in 3' $\rightarrow$ 5' direction (ascending) and another strand runs in 5' $\rightarrow$ 3' direction (descending).

Each nitrogen base in DNA is attached horizontally to carbon atom 1 of sugar molecule. The N-bases in DNA are paried and joined with each other by weak hydrogen bonds. This is called base-pairing. The base pairing is very specific and complementary to each other. The pairing occurs in between purine and pyrimidine nitrogen bases. Adenine always pairs with thymine with two hydrogen bonds (A = T) and Guanine always pairs with cytosine with three hydrogen bonds (G $\equiv$ C). Thus, the two strands of DNA molecule are not similar but complimentary to each other i.e. sequence of nucleotides in one strand can determine the sequence of nucleotides in other strand. Due to complimentary base pairing the purine and pyrimidine ratio in DNA is always 1 : 1.

$$A + G = C + T \text{ OR } \frac{A + G}{C + T} = 1$$

Chargaff in 1950 found that purines and pyrimidines occurs in equal amount in DNA molecule. A + G = C + T i.e. Chargaff rule. According to Chargaff, phosphate and sugar molecules occur in equal number. DNA shows two types of grooves namely major and minor groove. Major grooves are formed due to coiling of double helix and minor grooves are formed due to the twisting of two strands. The DNA molecule has three dimentional structure. The diameter of DNA is **20 A°**. Each complete spiral or turn of DNA strand has distance about **34A°** which consists of 10 base pairs. The distance between two successive base pairs is **3.4 A°** or **0.34 nm**. The base pairs are rotated by **36°** with respect to each adjacent pair, so that complete turn of **360°** involves 10 base pairs. The angle of one complete turn is **360°** and it is known as turn or pitch angle. The width of base pairs is **10.8 A°** to **11 A°**. Four different types of nucleotides are repeatedly present in DNA, hence it is called heteropolymer.

Biological Significance of DNA :

DNA is important molecule in the living body. It performs following function in the living world.

1. DNA contains genetic information in coded form. It carries genetic information from one generation to another generation by the property of replication.

2. It controls all metabolic activities directly or indirectly by synthesizing specific proteins.

3. It helps in synthesis of RNAs which carry out protein synthesis.

4. It plays an important role in production of variations through mutations which may lead to the process of evolution.

Forms of DNA :

DNA is a polymorphic molecule hence shows different structures or forms. The forms of DNA are A, B, C, D and Z. These forms of DNA are different from each other with respect to, the number of base pairs per turn, angle between the base pairs, diameter of DNA molecule and handedness of double helix i.e. right handed or left handed.

1. **B-DNA :** It is most common form of DNA. It is double stranded and right handed form. It has 10 base pairs per turn of helix at distance of 3.4 A°. It is proposed by Watson and Crick. The width of helix is 20 A°.

2. **A-DNA :** It is right handed double helix similar to the B-DNA. It has 11 base pairs per turn and a diameter of helix is 23 A°. It has smaller twist angle resulting in change in groove depth. Depth of major groove increased and smaller groove decreased.

3. **C-DNA :** It is right handed double helix with 9 base pairs per turn. The diameter of helix is 23.7A°.

4. **D-DNA :** It is also right handed form of DNA and is more or less similar to C-DNA. It has 8 base pair per turn. It is rarely found.

5. **Z-DNA :** It is left handed double helix model with zig-zag arrangement of sugar phosphate molecules in backbone. Running in antiparallel direction. Hence, it is termed as Z-DNA. It has 12 base pairs per turn with diameter 18A°. It is found in large number of living organisms including mammals, protozoans and several plant species.

1.5 RIBONUCLEIC ACID (RNA)

RNA is long, single stranded, high molecular weight polymer compound. It is composed of ribose sugar, nitrogen base and phosphate group. It is similar to DNA in its polynucleotide chain. The nucleotides present in RNA are ribonucleotides. It is present in the cytoplasm and nucleus of the cell. In chloroplast and mitochondria these are found in various shapes. RNA is present in all organisms as non-genetic materials except in some viruses where DNA is totally absent.

Chemically RNA is composed of 3 components, namely ribose sugar, phosphoric acid and nitrogen base. Sugar in RNA is pentose sugar i.e. ribose ($C_5H_{10}O_5$). Phosphoric acid molecule is with three reactive (–OH) groups and nitrogen bases are similar to DNA with one base change from DNA. In RNA uracil is present instead of thymine in DNA. Bases in RNA are adenine, guanine, cytosine and uracil.

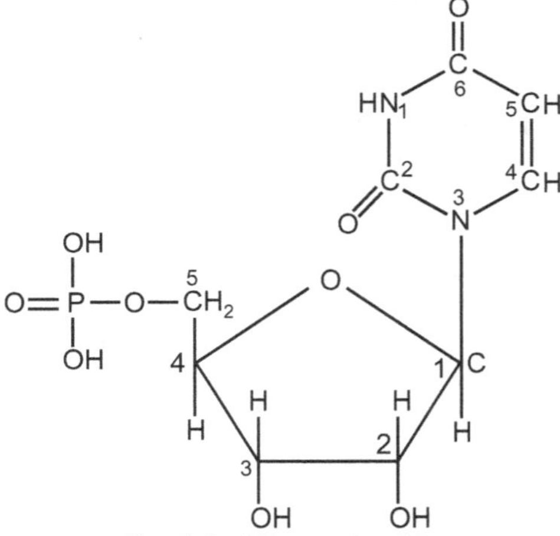

Fig. 1.8 : Ribose sugar

Fig. 1.9 : Ribonucleotide

1.6   STRUCTURE OF RNA

RNA is single stranded nucleic acid found in living organisms. Its molecular weight is high hence it is called macromolecule. RNA molecule is formed by many smaller units or monomers called ribonucleotides. Each ribonuleotide is made up of ribose sugar, phosphate group and nitrogen base. A nitrogen base is attached to carbon atom no. 1 of ribose sugar and phosphate group is attached at 5th carbon atom of ribose sugar to form nucleotide. Four types of nucleotides are present in RNA. These are, adenine, guanine, cytosine and uracil nucleotides. In RNA, ribonucleotides are joined together by phosphodiester linkages in a long chain to form single stranded RNA molecule. The RNA strand has two ends 3' end and 5' end. The strand of RNA may fold upon itself in certain regions to form loops. The loops contain bases in pairs i.e. A = U and G ≡ C. In unfolded or linear regions, the nitrogen bases remains unpaired. The base paring in RNA folds is necessary for stability of RNA molecule. Because of single stranded nature and base paring, RNA does not have 1:1 purine pyrimidine ratio.

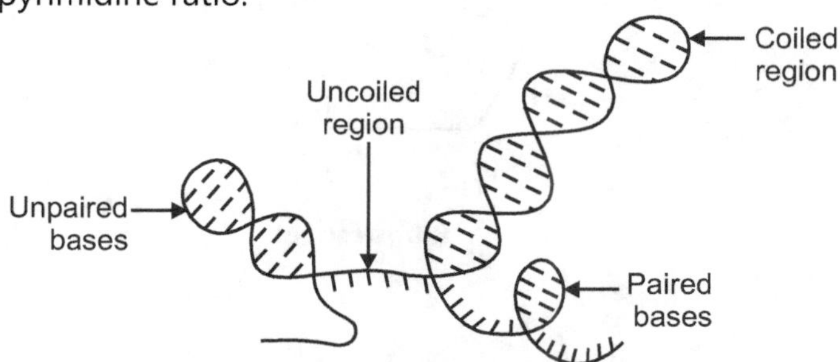

Fig. 1.10 : RNA Polynucleotide Chain

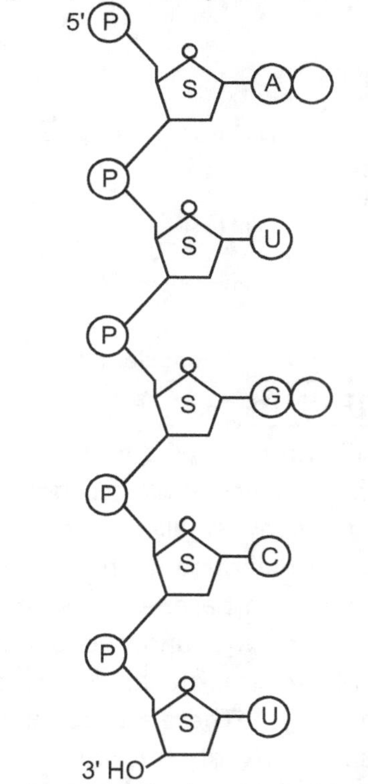

Fig. 1.11 : Single strand of RNA

Functions of RNA :

1. RNA acts as genetic material in many viruses.
2. Different types of cellular RNAs play important role in protein synthesis.
3. It also initiates the process of DNA replication.

Types of RNA :

Basically RNA is categorized into two types.

1. Genetic RNA.
2. Non-genetic or cellular RNA.

1. **Genetic RNA :** The RNA that carries genetic information and acts a genetic material is called genetic RNA. All plant viruses and few animal viruses have RNA as genetic material.

    e.g. Tobacco mosaic virus, Influenza virus etc.

2. **Non-genetic or Cellular RNA :** The RNA which does not acts as genetic material is called non-genetic or cellular RNA. Cellular RNAs help in protein synthesis and other functions of the cell. They are present in all organisms except in some viruses. All types of non-genetic RNA are synthesized on DNA templet by the process called transcription. On the basis of molecular structure and functions, RNA is classified into 3 types,

    (A) Messenger RNA (mRNA)

    (B) Ribosomal RNA (rRNA) and

    (C) Transfer RNA (tRNA)

1.6.1 Messenger RNA (mRNA)

It is single stranded, linear unfolded molecule. Hence, base pairing is totally absent. It forms only 3-5% of total cellular RNA.

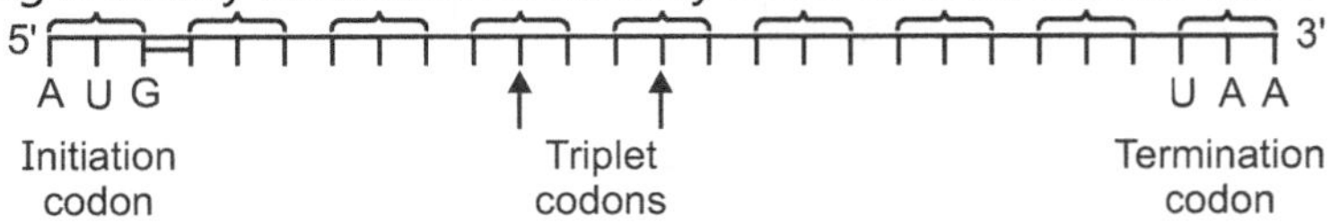

Fig. 1.12 : mRNA

As it carries information for synthesis of protein from DNA to ribosomes in cytoplasm hence it is called messenger RNA or mRNA (Jacob and Monod 1961). It is longest RNA with average molecular weight 5,00,000. It is produced on the DNA strand (templet) in the nucleus by the process called transcription and then transferred to cytoplasm i.e. at the site of protein synthesis. Hence, the sequence of nitrogen bases on the mRNA is complimentary to the sequence of DNA on which it is synthesized. The number of nitrogen bases varies from 900 to 12,000 based on the type of protein. The sequence of three successive bases on mRNA forms triplet called **codon**. Each codon specifies one amino acid. This coded language of mRNA is called mRNA language or **genetic code** or **cryptogram**. mRNA is never folded structure. It has direction from 5′ to 3′.

At 5′ end, the initiation or **starting codon** AUG or sometimes GUG is present. AUG always codes for amino acid methionine or f-methionine. At 3′ end the **stop** or **non-sense condons** either UAA, UAG or UGA are present which is last codon. They are also called **termination codon** as they terminate synthesis of polypeptide chain. The length of mRNA depends upon number of codons specifying the amino acid. mRNA is short lived due to its instability. In prokaryotes, it may live from one or two minutes while in eukaryotes it may last for many hours or upto even days. Each gene transcribes its own mRNA, therefore number of mRNAs are equal to number of genes in a cell. mRNA may be monocistronic i.e. made up of single gene or polycistronic made up of many genes. Monocistronic mRNA is present in eukaryotes and polycistronic mRNA is present in prokaryotes.

Structurally, mRNA molecule is composed of following regions,

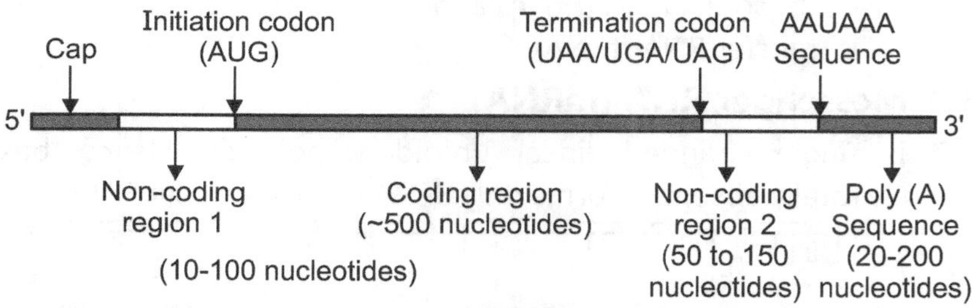

Fig. 1.13 : Structure of mRNA in eukaryotes

(a) Cap : It is the beginning of the mRNA at the 5′ end. It is found in eukaryotic cells and animals viruses. The cap of mRNA molecule decides the rate of protein synthesis. Cap helps in binding the mRNA molecule with the ribosome.

(b) Non-coding region 1 : The region next to cap is non-coding region 1. It is 10-100 nucleotide long and having more A and U residues. It is non-translating region.

(c) Initiation codon : It is the region of three nitrogen bases (triplet codon) AUG next to NC 1. It always codes for amino acid methionine which is starting amino acid in any protein molecule. Sometimes it is replaced by codon GUG.

(d) Coding region : It is main and functional region of mRNA. This region contain many nucleotides based on the size of protein molecule i.e. number of amino acids in the protein. Average number of nucleotides are 1500. The coding region varies in different mRNA based on genes.

(e) Termination codon : The coding region of mRNA is ended by the presence of codons such as UAA, UAG or UGA. These codons terminates the polypeptide chain (protein) synthesis hence called termination codons.

(f) Non-coding region 2 : It is non-coding region of 50-150 nucleotides. It does not translate proteins. It consist of shorts sequence of AAUAAA.

(g) Poly (A) sequence : At 3′ end of mRNA a sequence of pure adenine residues of 20-200 nucleotides. It is made up of many adenine residues. The polyadenylation of mRNA is carried out in the nucleus.

Functions of mRNA :

(1) It carries genetic information from DNA to ribosome during protein synthesis.

(2) The genetic code of mRNA get translated into the sequence of amino acids to form proteins.

1.6.2 Ribosomal RNA (rRNA)

This RNA is found in ribosomes or associated with ribosomes hence named ribosomal RNA. It constitute about 70-80% of total cellular RNA. It is a single stranded folded molecule. In the folded regions, the bases are paired. The base pairing is complimentary in folded regions and are joined by weak hydrogen bonds.

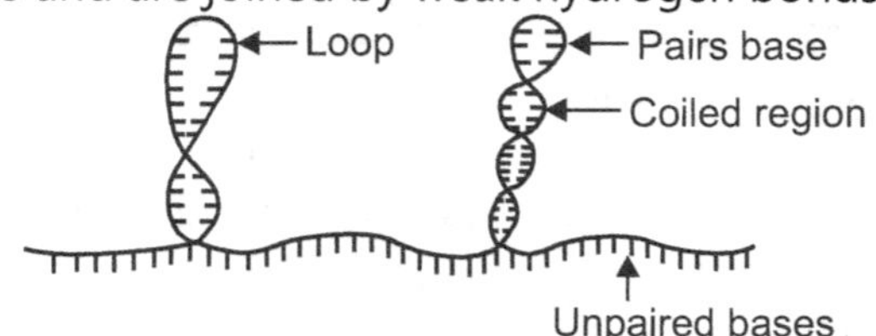

Fig. 1.14 : rRNA

In unfolded regions, the bases are unpaired hence rRNA does not show purine pyrimidine equality. The rRNA consists of two ends a 5′ phosphate end and 3′ OH end. It is the most stable kind of RNA and also known as insoluble RNA. It is stable at least for two generations. The strand folds and unfolds on cooling and heating respectively.

rRNA is synthesised on ribosomal DNA. In prokaryotes, it is synthesized in the cytoplasm while in eukaryotes, it is synthesized in nucleolar organizer present in nucleolus. The sequence of nitrogen bases of rRNA is complementary to the bases on DNA on which it is synthesized. The rRNA formed in nucleolar organizer is 45 S RNA. It then cleaved to 32 S and 18 S and finally 32 S RNA cleaved into 28 S. The 5 S RNA is transcribed by the genes present outside the nucleolar organizer.

rRNA shows variation in its forms as per the type of ribosome. In 70 S ribosomes present in prokaryotes 16 S, 23 S and 5 S RNAs are present. In 80 S ribosomes found in eukaryotes 18 S, 28 S and 5 S RNA and in mitochondrial ribosomes 12 S, 16 S and 5 S RNAs are present. The molecular weight of rRNA ranges from 40,000 to more than 1 million.

Functions of rRNA :

    (i)   It provides proper binding site to mRNA on the ribosomes.

    (ii) To produce the structural ribosome which is a site of protein synthesis.

    (iii) It helps mRNA to orient in such a way that its nitrogen bases are easily identified to read its all codons.

1.6.3 Transfer RNA (tRNA)

The RNA which transfers the activated amino acid from cytoplasm to the site of protein synthesis i.e. ribosomes is called transfer-RNA. It is also called soluble RNA (sRNA). It is about 10 to 20% of total RNA of the cell. tRNA is small, single stranded globular molecule containing about 73-93 nucleotides.

It has low molecular weight of about 23,000-30,000 Daltons. tRNA is synthesized in nucleus on DNA templet. The nitrogen bases present in tRNA are adenine, guainie, cytosine, uracil, pseudouridine (Ψ), dihydrouridine (DHU), inosine (I) and thymine. The unusual bases in tRNA are formed by methylation. tRNA shows specificity with amino acids and hence coded by separate genes.

The single strand of tRNA is looped about itself and shows two different structural models as :

    (1) Clover leaf model-(Trifoliate) and

    (2) Hairpin model.

1. **Clover-leaf Model :** This model of tRNA was proposed by Holley et. al. (1965). It is widely accepted model of tRNA. According to this model a single strand of tRNA is folded upon itself producing cloverleaf like or trifoliate structure. As a result of folding, the 3' and 5' end of chain come near to each other and the two free arms are unequal. The 3' end always terminates into CCA sequence to which amino acid is attached. Hence, called acceptor arm or carrier site. The 5' end terminates into G or C base.

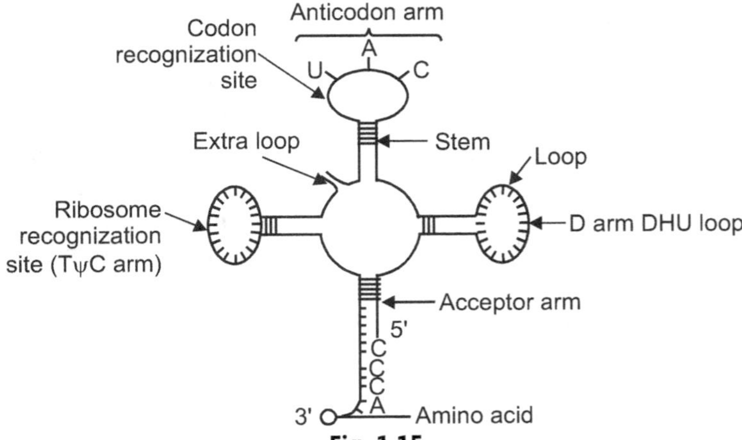

Fig. 1.15

According to this model, tRNA molecule consists of 3 major arms, one variable arm (Lump) and two free ends (3' and 5'). Each arm has a stem and terminal loop. The middle loop bears three unpaired nitrogen bases that are directed outside the loop, called as anticodon, which is complementary to codon on mRNA. The various arms in tRNA molecule are :

(i) **Acceptor arm :** It is opposite to anticodon arm. It consists of a stem of 7 paired bases and a terminal sequence of four unpaired nucleotides. It accepts amino acids. The loop is absent. The stem has two unequal ends – 3' and 5' end. The 3' end has CCA ending and 5' end have G or C unpaired base.

(ii) **T Ψ C or T-arm :** It has stem as well as loop. The stem with 4-5 paired nitrogen bases and loop with 7 nucleotides. The outermost pair of stem is C-G. It contain T Ψ C constant sequence in loop. The loop shows presence of unusual nitrogen base a pseudouridine (Ψ). It has ribosome recognition site or ribosome binding loop.

(iii) **DHU or D-arm :** It also has stem and loop in it. The stem contain 3-4 base pairs and loop has 7-11 unpaired nitrogen bases. The loop contain unusual base dihydrouridine (DHU). Hence, the loop is called DHU loop or D loop. It has binding site for amino acid activating enzyme (synthatase) for partial binding.

(iv) Variable arm : It is also called extra arm or minimum or lump. It is present between T Ψ C and anticodon arm. It may be of two types. One without stem and having only 4-5 bases and another with 13-21 bases having both stem and loop.

(v) Anticodon arm : It lies opposite to acceptor arm. It has arm with 5 paired nitrogen bases. The loop called anticodon loop has 7 unpaired bases. The middle 3 bases of loop forms anticodon which is complementary to codon on mRNA. It recognizes codon on mRNA hence, called codon recognition site. The anticodons are also called as **nodoc**. The loop has hypermodified purine (H), U and a pyrimidine (V).

2. **Hairpin Model :** This model was proposed by Hogland. In this model, the folding results into formation of one loop and two unequal free arms. The loop has a triplet of 3 unpaired nitrogen bases called as anticodon. The 5' end has G-base and 3' end has a sequence of CCA nitrogen bases.

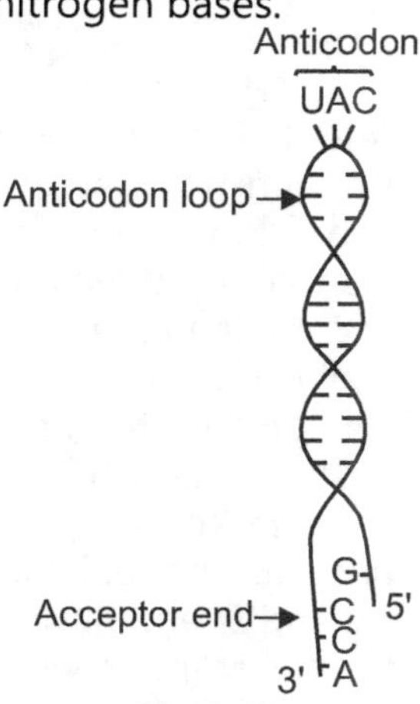

Fig. 1.16

Functions of tRNA :

(i) tRNA carries activated amino acids from the amino acid pool to site of protein synthesis i.e. ribosomes. Therefore called **adaptor** molecule.

(ii) It helps to arrange the amino acids in their proper sequence for the synthesis of proteins.

(iii) Specific amino acid is picked up by specific tRNA and carried it at 3' end having CCA sequence.

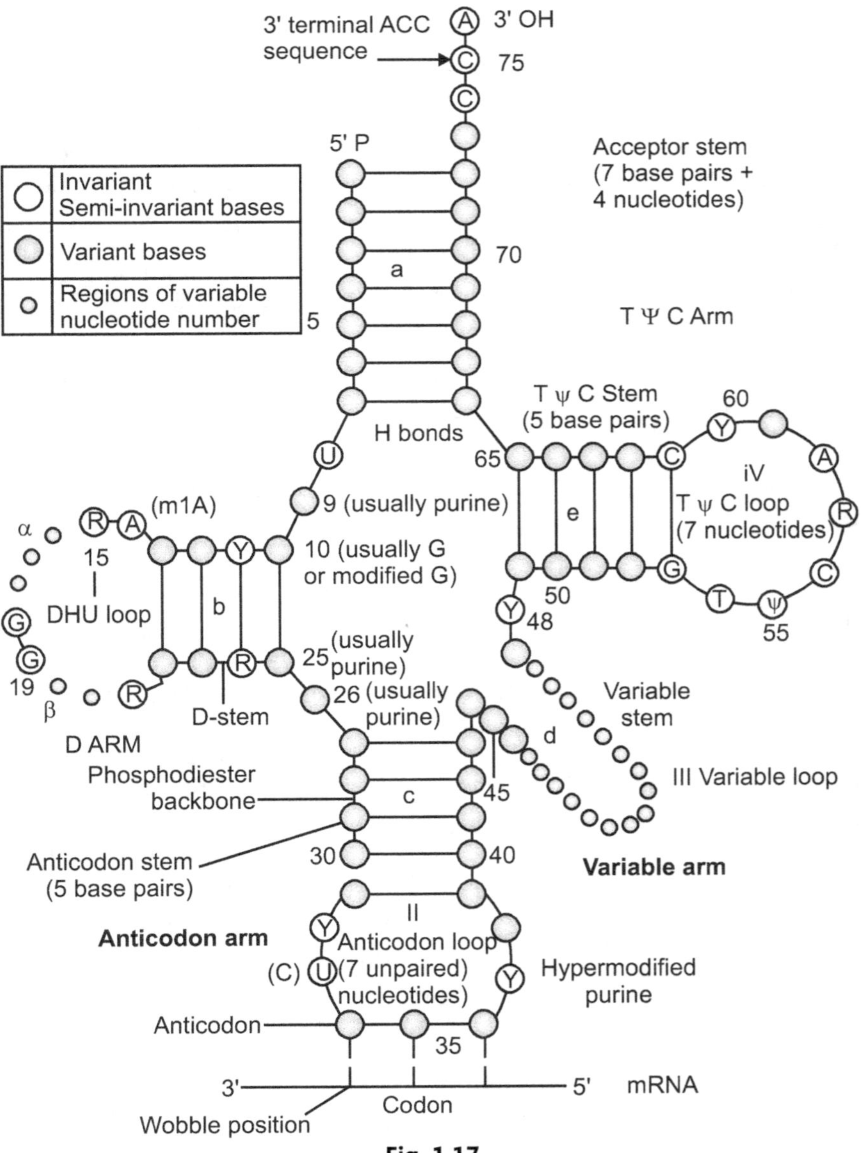

Fig. 1.17

Difference between RNA and DNA :

Sr. No.	RNA	DNA
1.	It is present in nucleus and cytoplasm.	It is present in Nucleus.
2.	RNA is single stranded.	DNA is double stranded
3.	The pentose sugar is ribose.	The pentose sugar is deoxyribose.
4.	Nitrogen bases are adenine, guanine cytosine and uracil (A, G, C, U)	The nitrogen bases are adenine, guanine, cytosine and thymine (A, G, C, T).
5.	Nitrogen base paired only in folded regions.	Nitrogen bases are paired throughout the length of DNA.
6.	Ratio of purine and pyrimidine is not 1 : 1.	Ratio of purine and pyrimidine bases is always 1 : 1.
7.	Transcribed on DNA molecule.	DNA is self replicating molecule.
8.	It helps in the process of protein synthesis.	It controls the process of protein synthesis.
9.	There are 3 types of RNA, mRNA, rRNA and tRNA.	DNA is of only one type.
10.	It contain less number of nucleotides i.e. up to 12000.	It contain large number of nucleotides i.e. up to 4.3 million.
11.	It has short life span and is replaced continuously.	It survives throughout the life span of the cell.
12.	The strand may fold to form loops.	The DNA is spirally twisted to produce double helix.
13.	The quantity of RNA depends upon the metabolic activity of cell.	The quantity of DNA is fixed for each cell of the individual.
14.	It often contains unusual nitrogen bases.	It does not contain unusual nitrogen bases.
15.	If forms the genetic material of some viruses only.	It is the only genetic materials of most of organisms.

EXERCISE

1. Explain deoxyribose nucleic acid.
2. Describe in detail structure of DNA
3. Describe watson and crick model of DNA.
4. Differentiate between RNA and DNA.
5. Describe in detail structure of RNA
6. Write short notes on the following:

(a) Nucleoside
(b) Nucleotide
(c) Significance of DNA
(d) Messenger RNA
(e) Ribosomal RNA
(f) Transfer RNA
(g) Clover-leaf Model
(h) Hairpin Model

CARBOHYDRATE METABOLISM

2.1 INTRODUCTION

In simple terms metabolism refers to all the enzymatic reactions that occurs in the cell. It involves all the changes occur from the entry of the nutrients up to the exit of the end products. Through this process energy is obtained and utilized for the growth, maintenance and performance of biological functions. Metabolism is distinguished into two types on the basis of its nature. Intermediary metabolism concern with metabolites and their inter conversions and energy metabolism concerned with energy production. Later is also termed as **bioenergetics**. Metabolism is divided into two major phases, namely anabolism and catabolism. Anabolism deals with enzymatic biosynthesis of macromolecules with expenditure of energy while catabolism deals with breakdown of complex nutrient molecules with release of energy in the form of ATP. Anabolism and catabolism occur concurrently and simultaneously in the cells but are independently regulated.

Carbohydrate Metabolism

Carbohydrates are optically active polyhydroxy aldehydes or ketones. Carbohydrates are classified into sugars and non-sugars. Sugars are sweet in taste and soluble in water. e.g. Monosaccharides and oligosaccharides. Non-sugars do not have sweet taste and they are insoluble in water. e.g. Polysaccharides. Carbohydrates are chief source of energy for living organisms. It forms the major bulk of human diet. They are mainly found in the form of polysaccharide such as starch and disaccharides such as sucrose, lactose and maltose. The carbohydrates are hydrolyzed by the digestive enzymes into monosaccharides like glucose, fructose and galactose in digestive tract. Monosaccharides absorbed by the intestine are transported to the liver through hepatic portal system. In liver, all the hexoses are converted into glucose. The glucose is finally utilized by the liver for various purposes. It may be utilized for obtaining energy, may be stored in the form of glycogen, and may be converted into fats or proteins or other carbohydrates. Glucose may be partially degraded under special circumstances into lactic acid by the process called Glycolysis.

2.2 GLYCOLYSIS

The term glycolysis is derived from two Greek words glycos-**sugar** and lysis–**breakdown.** Hence, the term glycolysis means "**Splitting of Sugar**". In glycolysis, glucose is stepwise broken down into pyruvic acid in the cytoplasm of the cell. It is anaerobic process. It is defined as, *'the process of conversion of a glucose into two pyruvic acid molecules with the help of glycolytic enzymes'*. It is also named as **Embden-Meyerhof-Parnas** pathway as it was investigated by these three scientists in 1930s. Glycolysis is primitive metabolic pathway operates in the cell. It helps the cell by providing the energy immediately from glucose. The overall reaction of glycolysis is,

$$C_6H_{12}O_6 + 2ADP + 2NAD + 2Pi \longrightarrow 2C_3H_4O_3 + 2ATP + 2NADH$$
$$\text{Pyruvic acid} + 2H^+ + 2H_2O$$

A glycolysis involves sequence of 10 enzyme catalyzed reactions. It occurs in the cytoplasm of the cell outside the mitochondria. The enzymes of glycolysis are found in the extra mitochondrial soluble fraction of the cell. The reactions in the glycolysis are :

1. **Phosphorylation :**

    In this reaction glucose is phosphorylated to glucose 6-phosphate with the help of enzyme *hexokinase* or *glucokinase*. This is an irreversible reaction which requires ATP and Mg^{++}.

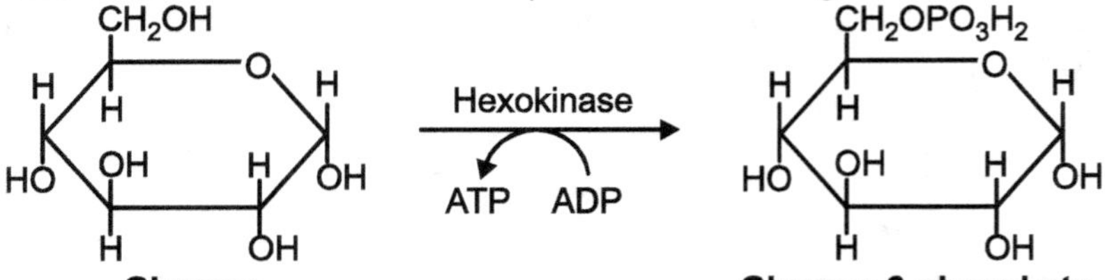

CH_2OH Hexokinase $CH_2OPO_3H_2$

ATP   ADP

Glucose   **Glucose 6-phosphate**

2. Isomerization :

Glucose 6-phosphate is converted into Fructose 6-phosphate in presence of enzyme *phospho glucose isomerase* in this reaction. This is reversible reaction.

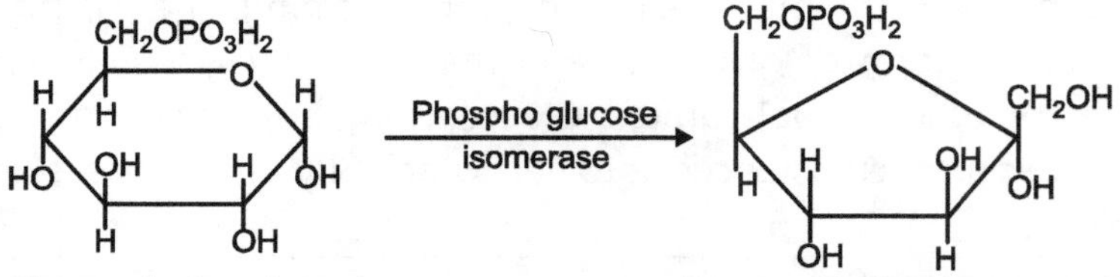

Glucose 6-phosphate   **Fructose 6-phosphate**

3. Phosphorylation :

Fructose is phosphorylated by ATP to Fructose 1, 6 –diphosphate in this reaction. The enzyme *phospho-fructo-kinase* catalyzes this reaction. This is second Phosphorylation reaction in glycolysis.

4. Cleavage of Fructose 1, 6 – diphosphate :

In this reaction, Fructose 1, 6 – diphosphate is broken down into two 3 carbon compounds namely glyceraldehydes 3-phosphate and dihydroxy acetone phosphate by the enzyme *aldolase*. This is also reversible reaction.

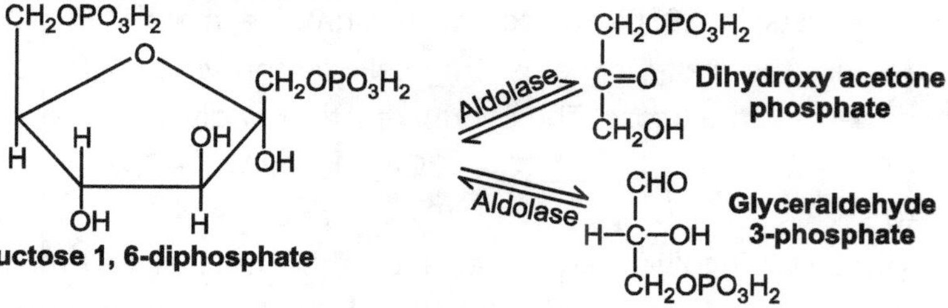

Fructose 1, 6-diphosphate

Dihydroxy acetone phosphate

Glyceraldehyde 3-phosphate

5. Isomerization :

The glyceraldehyde 3-phosphate formed by cleavage of Fructose 1, 6- diphosphate is directly enters the further pathway up to the formation of pyruvic acid. The dihydroxy acetone phosphate (DHAP) cannot entered directly, hence is isomerizes into glyceraldehyde 3-phosphate which is its isomer. The isomerization reaction is catalyzed by the enzyme *triose phosphate isomerase*. Finally from one molecule of Fructose 1, 6-diphosphate 2 molecules of glyceraldehyde 3-phosphate are formed.

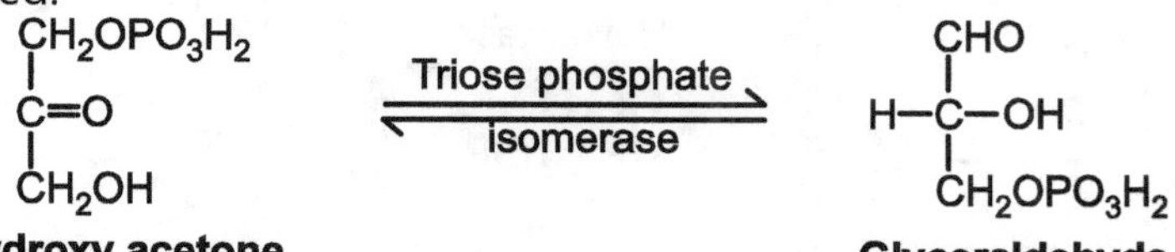

Dihydroxy acetone phosphate

Glyceraldehyde 3-phosphate

6. Oxidative Phosphorylation :

In this step, glyceraldehyde 3-phosphate is oxidized and phosphorylated simultaneously to form 1,3-diphosphoglyceric acid. The reaction is catalyzed by the enzyme *glyceraldehyde 3-phosphate dehydrogenase* in presence of NAD^+ and inorganic phosphate. The oxidized NAD is converted into $NADH_2$ which on electron transport chain generates only 2ATP molecules instead of 3ATP molecules. As $NADH_2$ is impermeable to mitochondrial membrane the electrons from $NADH_2$ are carried into mitochondria by **Glycerol phosphate shuttle**. In mitochondria these are used in synthesis of $FADH_2$.

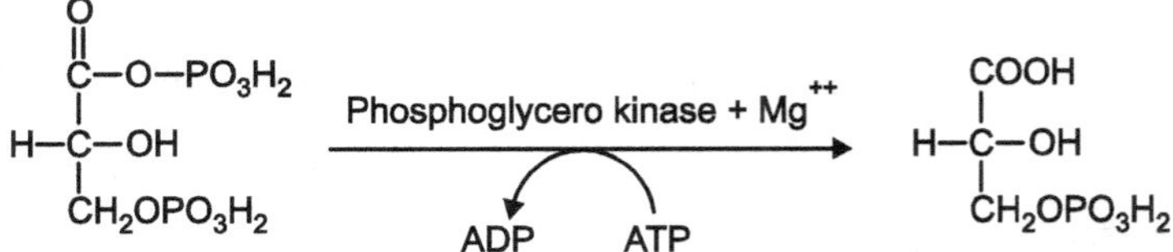

7. ATP generation or Dephosphoryation–I :

The high energy phosphate bond containing 1, 3-diphosphoglyceric acid undergo dephosphorylation to form 3-phosphoglyceric acid. The high energy phosphate group transferred is accepted by ADP to form ATP. The enzyme *phosphoglycerate kinase* in presence of Mg ++ catalyze the reaction. The phosphorylation in this reaction is called substrate level phosphorylation.

8. Isomerization (Rearrangement or position Isomerization):

In this reaction, internal arrangement of phosphate group between third C-atom and second C- atom of 3-phosphoglyceric acid occurs. The phosphate group is transferred from the third position (C_3) to the second position (C_2). This reaction is catalyzed by the enzyme *phosphoglyceromutase*.

9. Dehydration (Removal of water):

In this reaction 2-phosphoglyceric acid is dehydrated and gets converted into phosphoenol pyruvic acid (PEPA) by the action of the enzyme **enolase** in presence of Mg^{++} or Mn^{++}. One water molecule is removed in this step hence called dehydration reaction.

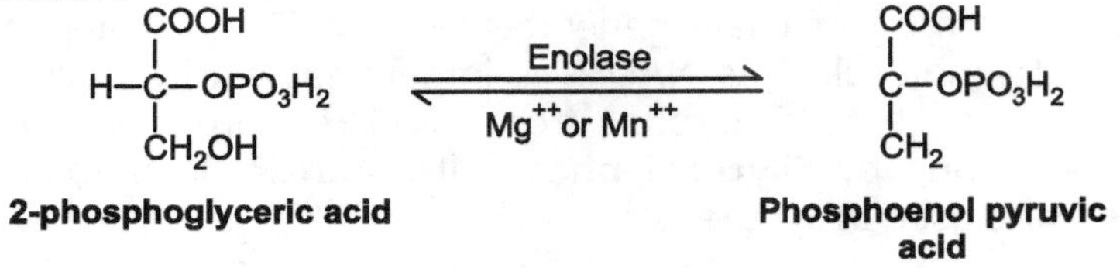

$$\underset{\textbf{2-phosphoglyceric acid}}{\begin{array}{c} COOH \\ | \\ H-C-OPO_3H_2 \\ | \\ CH_2OH \end{array}} \quad \underset{Mg^{++} \text{ or } Mn^{++}}{\overset{Enolase}{\rightleftharpoons}} \quad \underset{\substack{\textbf{Phosphoenol pyruvic} \\ \textbf{acid}}}{\begin{array}{c} COOH \\ | \\ C-OPO_3H_2 \\ | \\ CH_2 \end{array}}$$

10. ATP generation or Dephosphorylation -II :

This is final step of the glycolysis. In this step phosphoenol pyruvic acid (PEPA) is dephosphorylated to form pyruvic acid. The phosphate group is used in the synthesis of ATP molecule in presence of enzyme *pyruvate kinase*. On hydrolysis, PEPA loses the phosphate group which combines with ADP and forms ATP. The pyruvic acid formed is 3 carbon compound. It is also called as pyruvate ($CH_3COCOOH$).

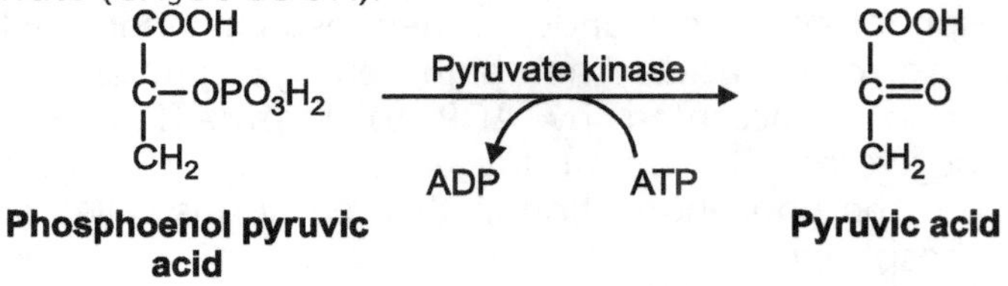

$$\underset{\substack{\textbf{Phosphoenol pyruvic} \\ \textbf{acid}}}{\begin{array}{c} COOH \\ | \\ C-OPO_3H_2 \\ | \\ CH_2 \end{array}} \quad \overset{\text{Pyruvate kinase}}{\underset{ADP \qquad ATP}{\longrightarrow}} \quad \underset{\textbf{Pyruvic acid}}{\begin{array}{c} COOH \\ | \\ C{=}O \\ | \\ CH_2 \end{array}}$$

Table 2.1 : Net gain of glycolysis

Step in Glycolysis	Product of reaction	ATP gain
Glucose ⟶ Glucose 6-phosphate	-1ATP	-1ATP
Fru 6-phos ⟶ Fru 1,6 diphosphate	-1ATP	-1ATP
1,3 DPGAL ⟶ 1,3 DPGA (2)	$2NADH_2$	6ATP
1,3 DPGA (2) ⟶ 3 PGA (2)	2ATP	2ATP
PEPA (2) ⟶ Pyruvic acid (2)	2ATP	2ATP
Net gain of ATPs		**8ATP**

Significance of Glycolysis :

1. In glycolysis glucose is incompletely oxidized and gives two molecules of pyruvic acid.

2. It generates 4 molecules of ATP directly through substrate phosphorylation and 6 ATP molecules by forming $2NADPH_2$.

3. Net gain of ATPs in the glycolysis is 8 ATP molecules.

4. Intermediate compounds generated in the glycolysis plays an important role in fat and protein metabolism.

5. There is no intake and direct use of O_2 and release of CO_2 in glycolysis.

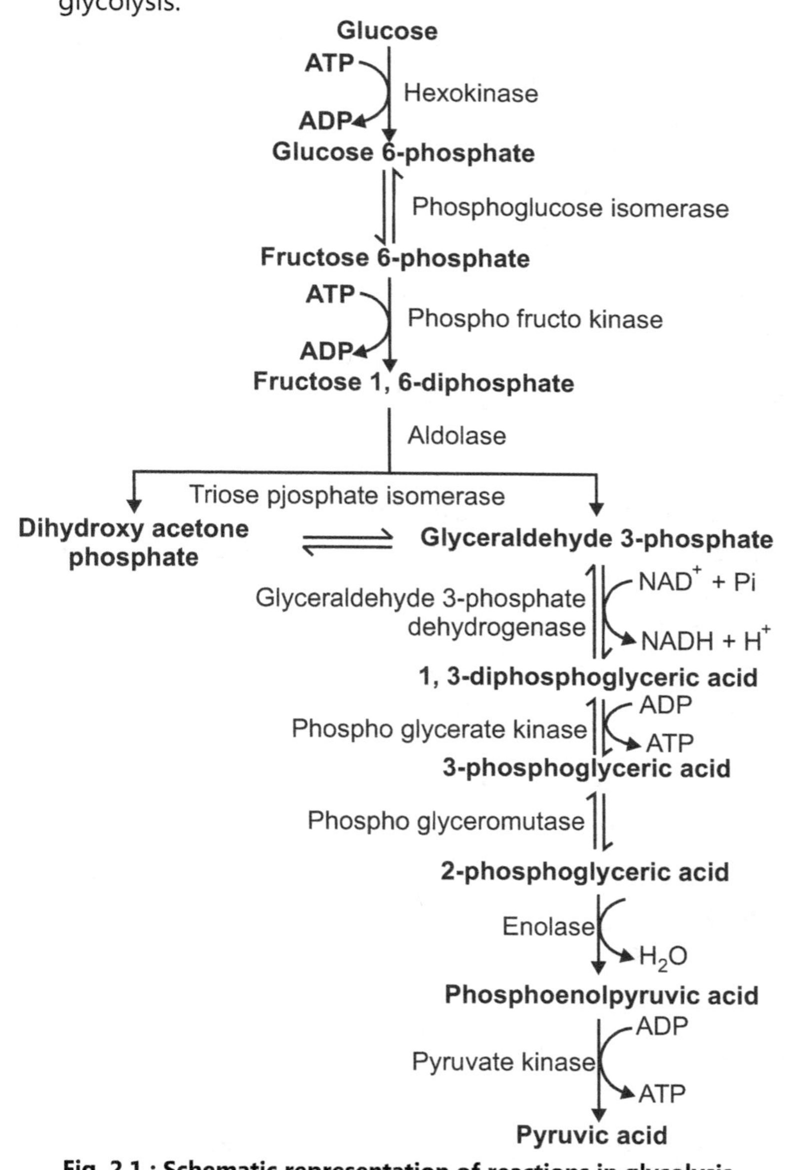

Fig. 2.1 : Schematic representation of reactions in glycolysis

2.3 KREB'S CYCLE

Kreb's cycle forms second major step or phase of aerobic respiration. It occurs in matrix of mitochondria. This is a central pathway for the relase of energy from acetyl CoA. The reactions involved in this cycle were first studied by sir Hans kreb in 1937 by using C^{14} isotope of carbon hence, it is named as Kreb's cycle. He received Nobel Prize in 1953 for this discovery of TCA cycle. He discovered this cycle in pigeon muscles. In this cycle, the first acid formed is citric acid hence the name given as **Citric Acid cycle**. The organic acids formed in the cycle are with three carboxylic groups (– COOH) hence, it is also named as tricarboxylic acid cycle or **TCA cycle**. In Kreb's cycle, Pyruvic acid is broken down into CO_2 and H_2O in the matrix of mitochondria in presence of molecular oxygen (O_2). The substrate entered is acetyl CoA and acceptor molecule is oxalic acid.

Steps involved in Kreb's Cycle :

After acetylation, the acetyl CoA enters into mitochondria and follows various reactions.

Step – 1 : Formation of Citric acid (Condensation) :

The cycle starts with the joining of two carbon Acetyl CoA with four carbon oxaloacetic acid to form six carbon citric acid. One molecule of H_2O is used in the reaction and CoA is released for recycling. This reaction is catalysed by the enzyme *citric acid synthatase*. Citric acid is the first stable compound of Kreb's cycle.

$$CH_3COSCOA + \underset{\substack{\text{Oxaloacetic}\\\text{acid}}}{\overset{\displaystyle O=C-COOH}{CH_2-COOH}} \xrightarrow{+\ H_2O} \underset{\text{Citric acid}}{\overset{\displaystyle COOH-CH_2}{HO-C-COOH}} + \underset{\text{Coenzyme-A}}{CoASH}$$

Acetyl Co-A

$$\text{Acetyl CoA} + \text{OAA} + H_2O \xrightarrow[\text{Synthase}]{\text{Citrate}} \text{Citric acid} + \text{CoA}$$

   (2C)   (4C)   (6C)

Step – 2 : Formation of Isocitric acid (Isomerisation) :

Citric acid is isomerized to isocitric acid.

It occurs in two steps :

(a) Dehydration : Citric acid is first converted to cis aconitic acid by removal of water.

(b) Hydration : The cis-aconitic acid is then converted into isocitric acid by addition of water.

These reactions are catalysed by the enzyme *aconitase*. It requires Fe^{++} as cofactor.

$$\text{Citric Acid} + H_2O \xrightarrow[\text{Aconitase}]{Fe^{++}} \text{Cis-aconitic acid}$$

(6C)   (6C)

$$\text{cis-aconitic Acid} + H_2O \longrightarrow \text{Isocitric Acid}$$

Step – 3 : Oxidation of Isocitric acid (Dehydrogenation - I) :

In this step the isocitric acid is oxidatively decarboxylated to α - Ketoglutaric acid in presence of enzyme *isocitrate dehydrogenase*.

First isocitric acid undergoes oxidation (dehydrogenation) to form oxalosuccinic acid with release of H_2. The pair of hydrogen atoms removed is accepted by NAD^+ and get reduced to $NADH_2$. Enzyme isocitrate dehydrogenase catalyse the reaction in presence of Mn^{++}.

$$\text{Isocitric acid} + \text{NAD} \xrightarrow[\text{Dehydrogenase}]{\text{Isocitrate } Mn^{++}} \text{oxalosuccinic} + NADH_2$$

(6C)   (6C)

Oxalosuccinic acid undergoes decarboxylation to form α - Ketogluatric acid (5C) with the removal of CO_2 in presence of enzyme *decarboxylase* and Mn^{++}.

$$\text{Oxalosuccinic acid} \xrightarrow[\text{decarboxylase}]{\text{Oxalosuccinate}} \alpha \text{ - ketoglutaric acid} + CO_2 \uparrow$$

(6C)   (5C)

Step – 4 : Second Oxydative decarboxylation (Dehyarogenation – II and Decarboxylation - II):

In this reaction α - ketoglutaric acid (5C) undergoes oxidation and decarboxylation to form succinyl Co-A (4C) by using CoA in presence of enzyme *dehydrogenase*. This reaction is similar to oxidative decarboxylation of pyruvic acid to Acetyl-CoA. This is an irreversible reaction. The reaction requires five cofactors as Mg^{++}, lipoic acid, NAD, CoA and decarboxylase.

$$\alpha \text{ - ketogutaric acid} + \text{Co-A} + \text{NAD (5C)} \xrightarrow[\text{Dehydrogenase}]{\alpha\text{-Ketoglutarate } Mn^{++}} \text{Succinyl-Co-A} + NADH_2 + CO_2$$

The hydrogen atoms released are accepted by NAD to form $NADH_2$.

Step – 5 : Phosphorylation and Hydrolysis (Formation of succinic acid):

Succinyl Co-A is converted into succinic acid (4C) by removal of Co-A and hydrolysis in presence of enzyme *succinyl thiokinase or succinyl CoA synthatase.* During this reaction, energy is liberated which is used in synthesis of GTP (Guanosine triphosphate) which reacts with ADP to form ATP by substrate level phosphorylation. This is the only reaction in the citric acid cycle to produce high energy phosphate bond directly.

$$\text{Succinyl CoA + GDP + iP + H}_2\text{O} \xrightleftharpoons{\text{Succinyl Thiokinase}} \text{Succinic Acid + GTP + CoA}$$

$$(4C) \hspace{6cm} (4C)$$

$$GTP + ADP \rightleftharpoons ATP + GDP$$

Step – 6 : Formation of Fumeric Acid (Dehydrogenation III)

Succinic acid (4C) is oxidized to fumeric acid (4C). The hydrogen released in reaction is accepted by FAD (Flavin Adenine Dinucleotide) and get reduced to $FADH_2$. The reaction is catalyzed by the enzyme *succinate dehydrogenase.* This is only reaction in kreb's cycle involving FAD as hydrogen acceptor. From $FADH_2$ only 2 ATP molecules are generated on electron transport chain.

$$\text{Succinic Acid + FAD} \xrightarrow[\text{Dehydrogenese}]{\text{Succinate}} \text{Fumeric acid + FADH}_2$$

$$(4C) \hspace{6cm} (4C)$$

Step – 7 : Formation of malic acid (Hydration) :

The fumeric acid by addition of water molecule get converted into malic acid (4C). The reaction is catalyzed by the enzyme *fumerase.*

$$\text{Fumeric acid + H}_2\text{O} \xrightarrow{\text{Fumerase}} \text{Malic acid}$$

$$(4C) \hspace{6cm} (4C)$$

Step – 8 : Oxidation of Malic acid (Dehydrogenation - IV) :

In the final step the malic acid (4C) is transformed into oxaloacetic acid (4C) which enters in kreb's Cycle again. Hydrogen released is accepted by NAD to form $NADH_2$. The reaction is catalyzed by the enzyme *malic acid dehydrogenase.*

$$\text{Malic acid + NAD} \xrightarrow[\text{Mg}^{++}, \text{Mn}^{++}]{\text{Dehydrogenase}} \text{Oxaloacetic acid + NADH}_2$$

$$(4C) \hspace{6cm} (4C)$$

Thus, each molecule of pyruvic acid in kreb's cycle forms 3 $NADH_2$, 1 $FADH_2$ and 1 ATP. The $NADH_2$ and $FADH_2$ enters the electron transport chain and produces 3 and 2 ATP molecules respectively.

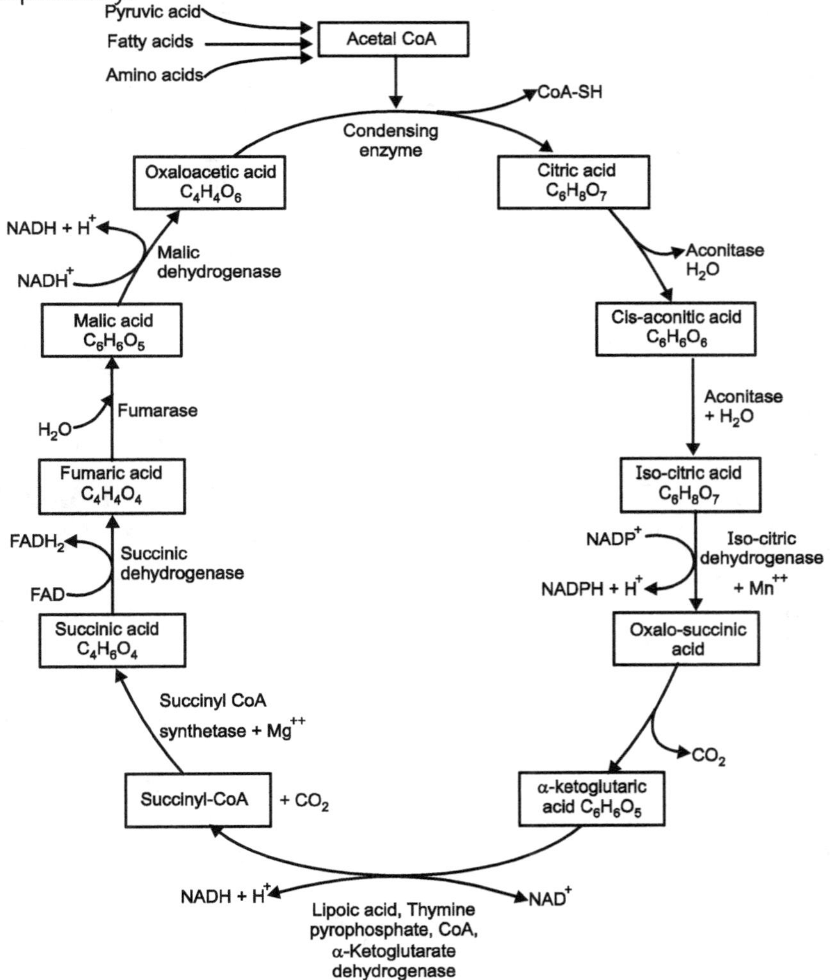

Fig. 2.2 : Kreb's OR TCA Cycle

Significance of Kreb's Cycle :

1. It is animportant exergonic cycle which brings about complete oxidation of pyruvic acid.

2. Oxidative phosphorylation of 2 molecules of pyruvic acid in Kreb's cycle generates 30 ATP molecules.

3. If provides intermediates for synthesis of fatty acids, glycerol, proteins, chlorophylls etc.

4. It helps in generating energy from fatty acids when carbohydrates are exhausted.

Table 2.2 : Net ATP gain from glucose from Kreb's cycle

Sr. No.	Step in Kreb's Cycle	ATP generated
1.	2 Isocitirc acid $\longrightarrow$ 2 Oxalosuccinic acid + $2NADH_2$ $2NADH_2 = (2 \times 3\ ATP)$	6ATP
2.	2 α - Ketoglutaric acid $\longrightarrow$ Succinyl CoA + 2 $NADH_2$ $2NADH_2 = (2 \times 3\ ATP)$	6ATP
3.	2 Succinly CoA $\longrightarrow$ 2 Succinic acid + 2 GTP $\rightarrow$ 2 ATP	2ATP
4.	2 Succinic Acid $\longrightarrow$ 2 Fumeric acid + $FADH_2$ $2FADH_2 = (2 \times 2\ ATP)$	4ATP
5.	2 Malic Acid $\longrightarrow$ 2 Oxalosuccinic Acid + 2 $NADH_2$ $2NADH_2 = (2 \times 3\ ATP)$	6ATP
	Net gain of ATPs	**24 ATPs**

2.4 PENTOSE PHOSPHATE PATHWAY: [HEXOSE MONOPHOSPHATE SHUNT (HMP - PATHWAY)]

The pentose phosphate pathway is also called the phosphoglyconate pathway. It is a metabolic pathway alternative or parallel to glycolysis. It is a aerobic pathway utilizing glucose as raw materials. It generates NADPH and pentoses (5C) as well as ribose 5-phosphate. The ribose 5 –phosphate is used in synthesis of nucleic acids. This pathway occurs in extra mitochondrial soluble portion of cells of tissues such as liver, adrenal cortex, thyroid, adipose tissue, testies, erythrocytes and lactating mammary glands. It is special because no energy in the form of ATP is produced or used up in this pathway.

Pentose phosphate pathway consist of two distinct phases :

(1) The oxidative phase and

(2) The non-oxidative phase

In oxidative phase conversion of hexose to pentose occurs while in non-oxidative phase pentose to hexose conversion occur. In the HMP, 5, 6, 3, 4, and 7 carbon sugars are formed.

(1) Oxidative Phase : In this phase oxidation of molecules occur. In oxidation phase, breakdown of molecules with loss of at least one electron takes place. The steps of this phase are ;

Step 1 : Phosphorylation : In this step the glucose is phophorylated to Glucose-6-phosphate by ATP in presence of enzyme *Hexokinase.*

$$\text{Glucose} \xrightarrow{\text{Hexokinase}} \text{Glucose-6-phosphate}$$

ATP   ADP

Step 2 : In this step Glucose-6-phosphate is oxidized to 6-phosphogluconolactone. The reaction is catalysed by the enzyme *glucose-6-phosphate dehydrogenase.*

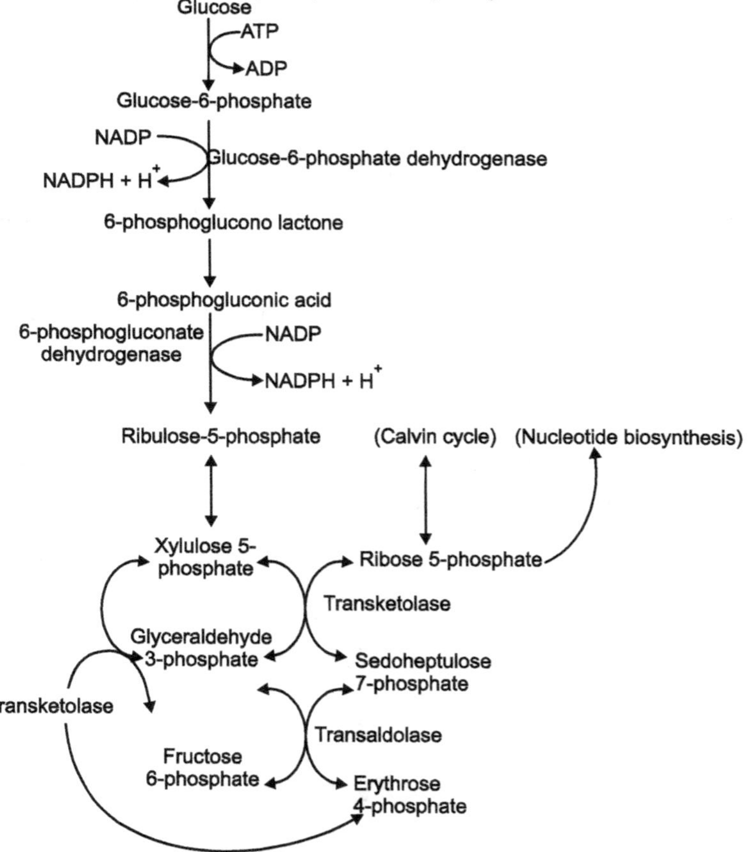

Fig. 2.3 : Pentose phosphate pathway

It requires NADP as reducing agent. The product formed is cyclic ester also called lactone.

$$\text{Glucose-6-phosphate} \longrightarrow \text{6-phosphogluconolactone}$$

NADP   NADPH + H⁺

Step 3 : The product of second step i.e. 6-phosphogluconolactone is unstable. It spontaneously hydrolyses to 6-phosphogluconic acid. The reaction is catalysed by the enzyme *gluconolactonase.*

$$\text{6-phosphogluconolactone} \xrightarrow[\text{+ H}_2\text{O}]{\text{Gluconolactonase}} \text{6-phosphogluconic acid}$$

Step 4 : In this step, 6-phosphogluconic acid is further oxidized to Ribulose 5-phosphate in presence of NADP. During the reaction unstable intermediate is formed. The intermediate loses CO_2 to give the pentose, ribulose 5-phosphate. The reaction is catalyzed by the enzyme *6-phosphogluconate dehydrogenase.*

$$\text{6-phosphogluconic acid} \xrightarrow[\text{NADP} \quad \text{NADPH + H}^+]{\substack{\text{6-phosphogluconate} \\ \text{dehydrogenase}}} \text{Ribulose 5-phosphate}$$

Step 5 : Ribose 5-phosphate is either converted into xylulose 5-phosphate or ribose 5-phosphate depending on the enzyme which catalyze the reaction. If *ribulose 5-phosphate epimerase* is the enzyme, the product is xylulose 5-phosphate and if *ribose 5-phosphate isomerase* is the enzyme, then the product will be Ribose 5-phosphate.

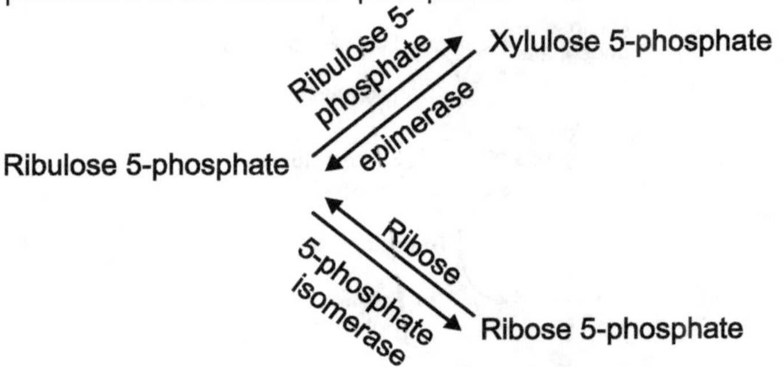

The above five steps constitute the oxidative phase of HMP shunt. In this phase, hexose molecule is converted into pentose molecule.

(2) The non-oxidative Phase :

In this phase, the 5 carbon sugars, by a series of reversible reactions get converted into fructose 6-phosphate and glyceraldehydes 3-phosphate, the glycolytic intermediates. The Ribulose 5-phosphate is the precursor to the sugar that makes up DNA and RNA. The steps in the phase are :

Step 6 :

The Xylusose 5-phosphate and Ribulose 5-phosphate reacts to form sedoheptulose 7-phosphate and *glyceraldehyde* 3-phosphate i.e. one heptose phosphate (C_7) and one triose phosphate (C_3) molecules. The reaction is catalysed by the enzyme *transketolase* :

Xylulose 5-phosphate	+	Ribulose 5-phosphate	$\xrightarrow{\text{Transketolase}}$	Sedoheptulose 7-phosphate	+	Glyceraldehyde 3-phosphate

Step 7 :

Sedoheptulose 7-phosphate reacts with glyceraldehyde 3-phosphate to form *erythrose* 4-phosphate (C_4) and fructose 6-phosphate (C_6) molecules. The reaction is catalysed by the enzyme *transaldolase*.

Sedoheptulose 7-phosphate	+	Glyceraldehyde 3-phosphate	$\xrightarrow{\text{Trans aldelose}}$	Erythrose 4-phosphate	+	Fructose 3-phosphate

Step 8 :

In this step, the erythrose 4-phosphate and xylulose 5-phosphate reacts with each other to form fructose 6-phosphate and glyceraldehydes 3-phosphate. The reaction is catalyzed by enzyme *transketolase*. These two molecules are responsible for the linking of HMP pathway with EMP pathway. By the reversible reactions glucose-6-phosphate nad glyceraldehydes 3-phosphate serves as common intermediate between these two pathways.

In humans and mammals, the pentose phosphate pathway occurs exclusively in the cytoplasm of cells, and is found most active in liver, mammary gland and adrenal cortex.

Significance of HMP shunt :

1. It produces reducing power namely NADPH. The NADPH is used for the synthesis of fatty acids, steroids, ascorbic acid etc. Production of NADPH is not linked to ATP generation in pentose phosphate pathway.

2. It provides ribose sugar for the synthesis of nucleic acids. Of the various pentoses formed, ribose-5-phsophate is used up in the biosynthesis of nucleic acids.

3. The products such as fructose 6-phosphate and glyceraldehydes 3-phosphate are used in glycolytic pathway.

4. The NADPH produced during HMP pathway may also be oxidized by electron transport chain, thereby each molecule of NADPH may generates 3 molecules of ATPs.

5. The erythrose 4-phosphate molecules produced in HMP by its conversion into shikimic acid and is indirectly acts as precursor of aromatic ring compounds.

2.5 GLUCONEOGENESIS (NEOGLUCOGENESIS)

In mammals, some tissues depends almost completely on glucose for their metabolic energy. The glucose requirements of the human brain are relatively enormous-about 120 gm per day and the amount of glucose generated from the body's glycogen reserves at any time is about 190 gm. In the conditions such as starvation for long time or diet with less amount of carbohydrates or intake of protein rich diet, the glucose deficiency occurs. The metabolic pathways other than glycolysis operates to synthesize the glucose from non-carbohydrate sources in liver and in cortex of kidneys. Very little gluconeogensis takes place in the brain, skeletal muscle and heart muscles. The pathways by which the synthesis of glucose from non-carbohydrate precursors occurs comes under gluconeogenesis.

The gluconeogensis is defined as, *"the synthesis of glucose from non-carbohydrate precursors (3-C, 4-C) is called gluconeogensis."*

The mechanism involved in gluconeogenesis is reversal of citric acid cycle and glycolysis. The substances forming intermediates of TCA cycle or glycolysis leads to glucose formation. The important substance are glucogenic amino acids, lactic acid, propionic acid and glycerol etc. In gluconeogenesis pyruvic acid is converted to glucose. However, gluconeogenesis is not an exact reversal of glycolysis. In glycolysis, there are three irreversible steps. These are,

1. Glucose + ATP $\xrightarrow{\text{Hexokinase}}$ Glu-6-phosphate + ADP

2. Fructose-6-phophate $\xrightarrow[\text{Kinase}]{\text{Phospho fructo}}$ fructose 1, 6-disphosphate + ADP

3. Phosphoenol pyruvate + ADP $\xrightarrow{\text{Pyruvate Kinase}}$ Pyruvic acid + ATP

In gluconeogenesis, these three reactions are by passed or substituted by the other ones. These are shown in the following Fig. 2.4.

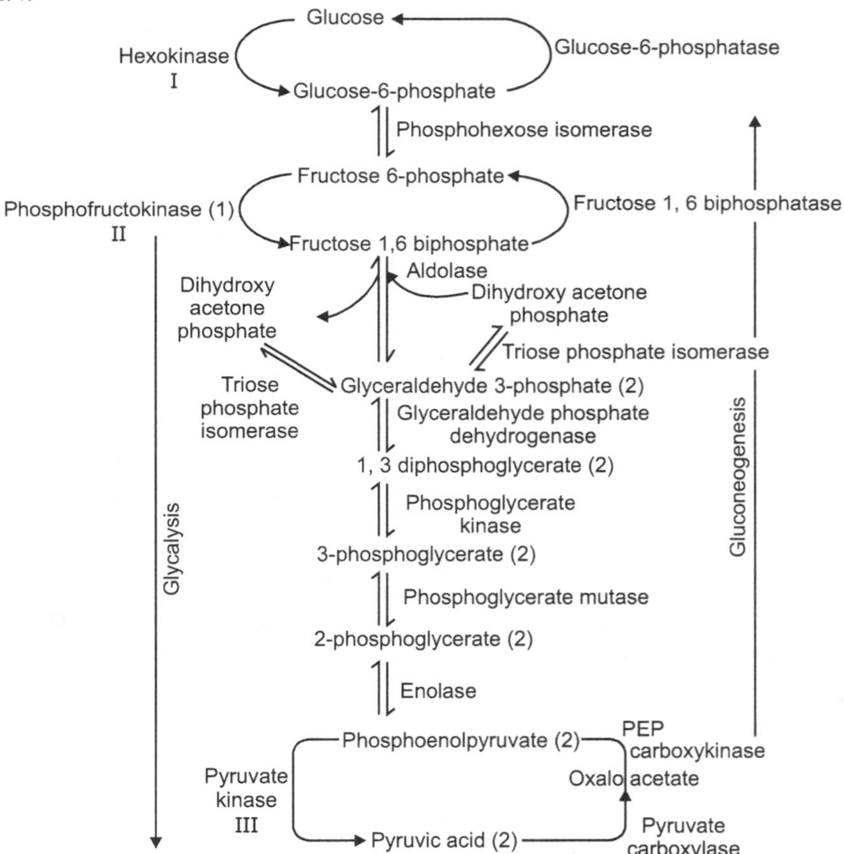

Fig. 2.4 : Glycolysis and Gluconeogenesis with 3 bypass reactions

(I,II or III)

Reactions are,

1. In the first reaction, pyruvic acid is carboylated to oxaloacetate at the expense of ATP. The reaction is catalyzed by pyruvate decarboxylase. The oxaloactate is then decarboxylated and phosphorlyated to form phosphoenol pyruvate with the expense of GTP. The reaction is catalyzed by phosphenol pyruvate carboxylase.

$$\text{Pyruvate} + CO_2 + ATP + H_2O \xrightleftharpoons[\quad]{\text{Pyruvate Carboxylase}} \text{Oxaloacetate} + ADP + Pi + 2H^+$$

$$\text{Oxaloacetate} + GTP \xrightleftharpoons[\text{Pyruvate Carboxylase}]{\text{Phosphenol}} \text{Phosphoenol Pyruvate} + GDP + CO_2 \uparrow$$

2. Fructose 1, 6 diphosphate gives fructose 6-phosphate on hydrolysis. The reaction is catalyzed by the enzyme fructose 1, 6 diphosphates.

$$\text{Fructose 1, 6 diphosphate} + H_2O \xrightarrow{\substack{\text{Fructose 1, 6}\\\text{diphosphatase}}} \text{Fructose 6-phosphate}$$

3. In the last reaction the fructose 6-phosphate is converted into glucose on hydrolysis. This reaction is catalyzed by the enzyme glucose 6-phosphatase.

$$\text{Glucose-6-phosphate} + H_2O \xrightarrow{\substack{\text{Glucose}\\\text{6-Phosphatase}}} \text{Glucose} + Pi$$

For each molecule of glucose formed from the pyruvic acid, six high energy phosphate groups (4 ATPs and 2 GTPs) are consumed. The extra price of gluconeogenesis is four high energy phosphate bonds per glucose synthesized.

Gluconeogenesis of Amino Acids and Propionic Acid :

The amino acids used in the synthesis of glucose are called glucogenic amino acids. By the process of transamination or deamination most of the glucogenic amino acids are converted into intermediates of Kreb's cycle. There are 20 types of amino acids in the cells. All these amino acids enter the kreb's cycle or TCA cycle by producing either one intermediate e.g. Alanine and glutamic acid or two intermediates e.g. phenylalanine and leucine. The amino acids are converted to intermediates like α-ketoglutarate, succinyl CoA, fumeric acid and oxaloacetic acid. Finally, the amino acids are metabolically routed through oxalacetic acid and phosphoenol pyruvic acid in the synthesis of glucose in gluconeogenesis. The propionic acid in ruminants is converted to succinyl CoA in TCA and is finally utilized in the synthesis of glucose by gluconeogenesis. The propionic acid is a saturated fatty acid formed as a product of carbohydrate fermentation in the rumen of ruminants.

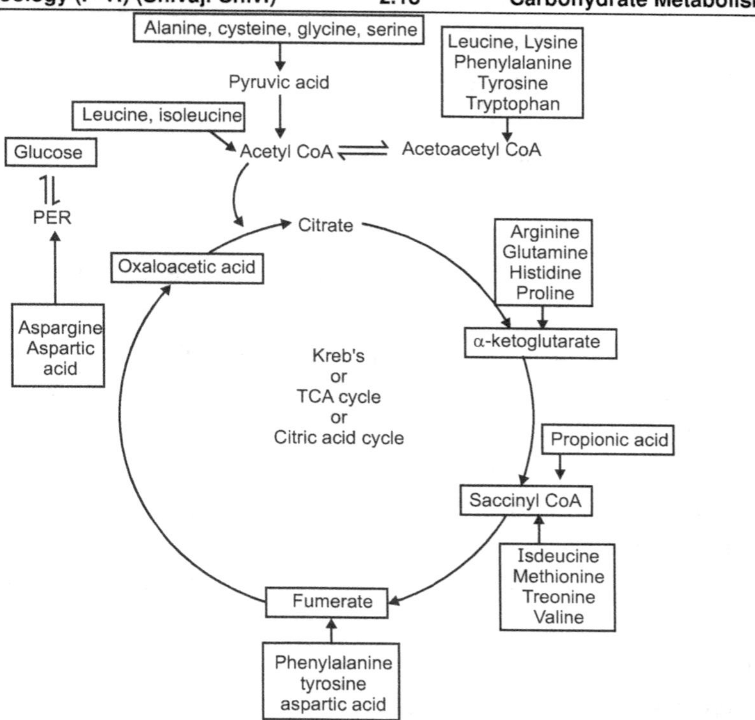

Fig. 2.5 : Entry of amino acids and propionic acid in TCA cycle

Gluconeogenesis of Lactic Acid (Cori cycle) :

It is the way of synthesis of glucose from lactic acid. The lactic acid is generated in the muscles during the heavy muscular work or straneous exercise due to inadequate oxygen supply. Hence, glycogen stored in the muscle get converted into lactic acid by glycogenolysis followed by anaerobic respiration i.e. glycolysis.

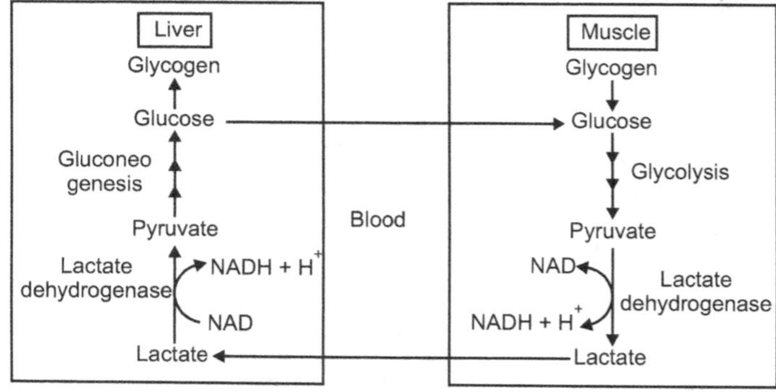

Fig. 2.6 : The Cori Cycle

The lactic acid accumulated in the muscle as muscle lacks enzyme glucose 6-phosphatase required for its conversions into glucose. The conversion of lactic acid into glucose takes place in the liver only.

The lactic acid enters the blood by diffusion and carried towards liver. In liver, lactic acid is converted to pyruvic acid by oxidation. Pyruvic acid is then converted into glucose by the process of gluconeogenesis in liver. The glucose is then used in the synthesis of glycogen in liver. The liver glycogen is brokendown into glucose again by the process of glycogenolysis and may be recycle to the muscle through blood. In muscle, the glucose enters and is used in the synthesis of muscle glycongen by the process called glycogenesis and thus the cycle completes. Hence, this is the cycle where liver glycogen may be converted into muscle glycogen and vice versa. The major raw material in the cycle is lactic acid. As the events occurs in cyclic manner it is called cycle and as it is invented by the scientist Kori, it is named as Kori cycle. Carl and Gerty Kori received Nobel Prize in 1937 for this work.

Gluconeogenesis of Glycerol :

During starvation for the gluconeogenesis another non-carbohydrate resource is made available by the fatty tissue. In adipose tissue, triglycerides are hydrolyzed and glycerol released. Glycerol formed is not metabolized further in adipose tissue due to lack of enzyme glycerokinase in it. The glycerol is then carried to liver where it is phosphorylated to glycerol 3-phosphate by the enzyme glycerokinase.

$$\text{Glycerol + ATP} \xrightarrow{\text{Glycero Kinase}} \text{Glycerol 3-phosphate + ADP}$$

Glycerol 3-phosphate is then oxidized to dihydroxy acetone phosphate (DHAP). The reaction is catalyzed by the enzyme glycerol phosphate dehydrogenase.

$$\text{Glucerol-3-phosphate + NAD}^{+} \xrightarrow{\text{Glycerol Phosphate dehydrogenase}} \text{DHAP + NADH + H}^{+}$$

The DHAP by its route though fructose 1, 6 diphosphate get converted into glucose.

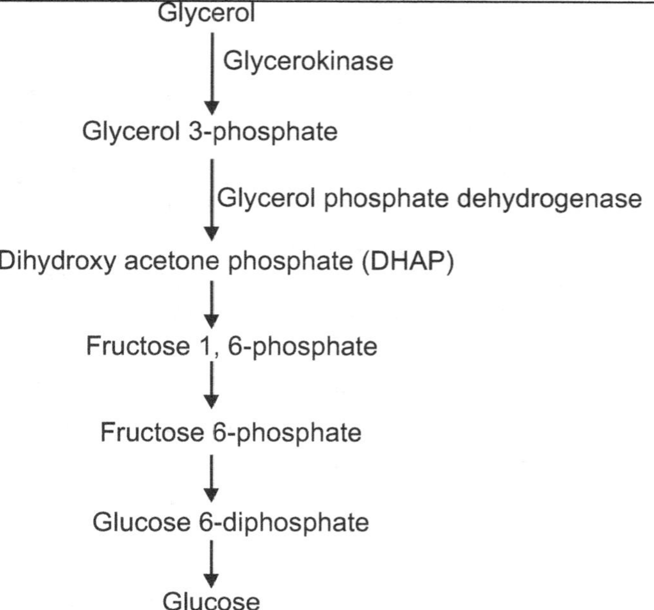

Fig. 2.7 : Glycerol Conversion into glucose

Significance of Gluconeogenesis :

1. It meets the need of the body for glucose when sufficient carbohydrate is not available from the diet or glycogen reserve.

2. It ensures the continuous supply of glucose to the organs such as brain and cells such as erythrocytes.

3. It helps in maintaining blood glucose level when fluctuate due to any reason.

4. Gluconeogenesis allows the use of dietary proteins in carbohydrate pathway.

5. The intermediates generated in gluconeogenesis helps in keeping the pathways active and operative as per the need of the cells.

2.6 GLYCOGENOLYSIS

Glycogen is the major storage form of carbohydrate in animals. It is a homopolymer made up of repeated units of α - D glucose and each molecule is linked to another by $1 \rightarrow 4$ glycosidic bond. It links first C-atom of the active glucose residue to the 6th C-atom of the approaching glucose molecule. The branching in glycogen begins by 1-6 linkages. The process of breakdown of glycogen is called glycogenolysis. Glycogen branches are catabolized by the sequential removal of glucose monomers by the enzyme glycogen phosphorylase. The glycogen stored in the liver, kidney and muscle cells of animals is broken down into glucose to provide immediate energy and to maintain blood glucose level during fasting or when sugar level in blood falls. It takes place in response to hormonal and neural signals. Glycogen may be broken down to glucose in liver and kidney or it may be broken down to glucose 6-phosphate in the muscles. The steps in the process of glycogenolysis are;

Step 1 :

In the 1st step, the glycogen is broken down into glycogen with one less glucose residue and glucose-1-phosphate. This reaction is catalysed by two independent enzymes, namely glycogen phosphorylase and debranching enzyme (α-1, 6 glucosidase).

Glycogen phosphorylase cleaves the bond linking a terminal glucose residue to a glycongen branch by substitution of a phosphoryl group for the α-1, 4 linkage. The removal of glucose residues as glucose 1-phosphate continue until about last 4 glucose residues on either side of the α-1, 6 branch. The chain remains with only four (4) glucose units is called **"Limit dextrin"**. The cleavage is known as **phosphorolysis**. The limit dextrin cannot broken down further by glycogen phosphorlyase.

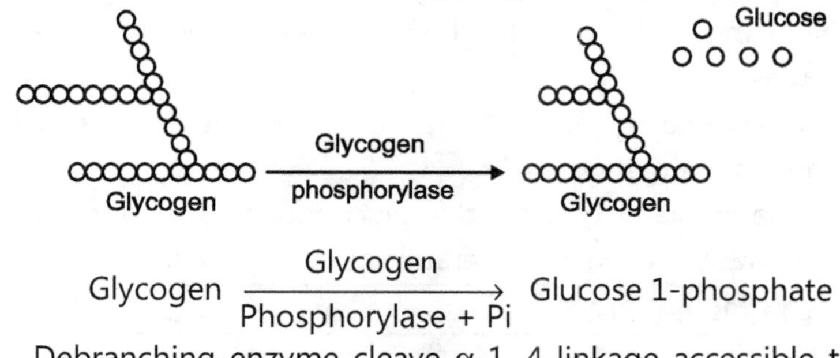

$$\text{Glycogen} \xrightarrow{\substack{\text{Glycogen} \\ \text{Phosphorylase + Pi}}} \text{Glucose 1-phosphate}$$

Debranching enzyme cleave α-1, 4 linkage accessible to the action of glycogen phosphorylase.

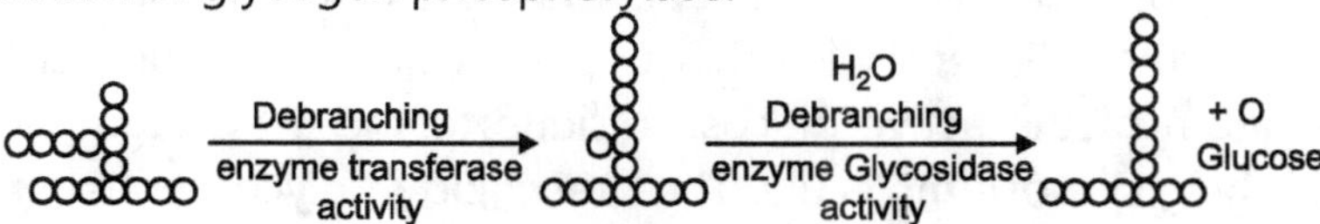

Fig. 2.8 : Action of Debranching enzyme

Thus by combined action of these enzymes the glycogen is converted to glucose 1-phosphate. The enzymes phosphorylase found in liver cells and muscle cells are different. In both organs active and inactive forms of enzyme are present. In presence of ATP the inactive form can be converted into active form.

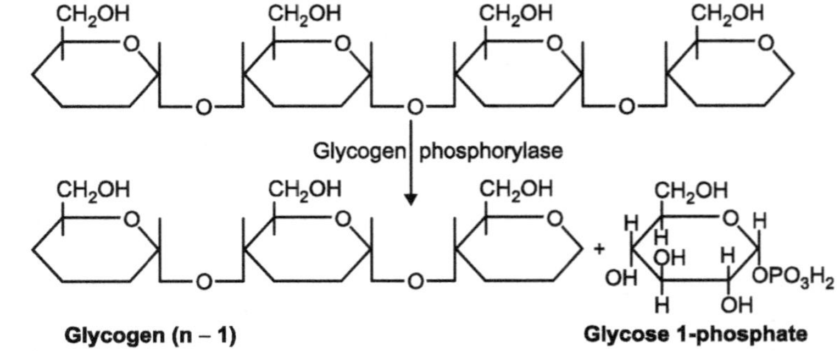

Glycogen phosphorylase

Glycogen (n – 1)   **Glycose 1-phosphate**

Step 2 :

The glucose 1-phosphate cleaved from glycogen is then reversibly converted into glucose 6-phosphate. The reaction is catalyzed by the enzyme phospho-glucomutase.

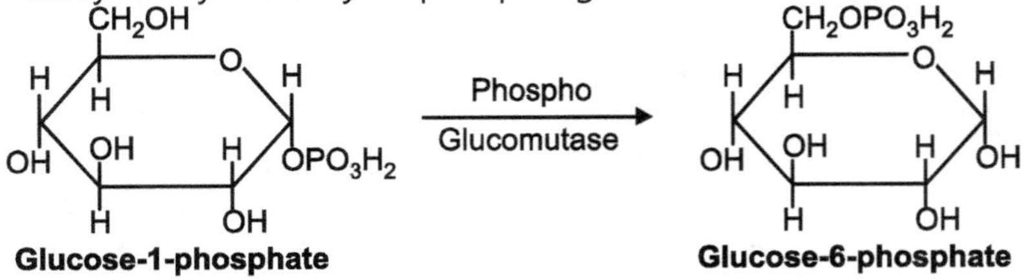

Phospho Glucomutase

Glucose-1-phosphate   **Glucose-6-phosphate**

Step 3 :

In this step glucose 6-phosphate is converted into glucose. The enzyme glucose 6-phosphatase catalyze the reaction. The enzyme removes the phosphate group from glucose 6-phosphate. This reaction takes place in the liver and kidney. It never occurs in muscle cells because muscles lack the enzyme glucose 6-phosphase.

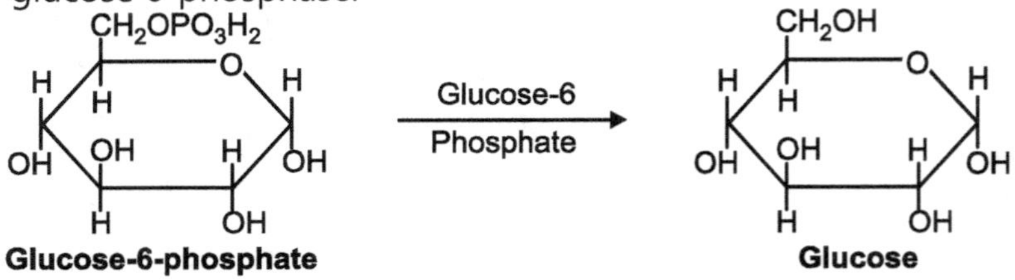

Glucose-6 Phosphate

Glucose-6-phosphate   **Glucose**

The glucose released from the glycogen after lysis is either utilized as energy source for carrying out cellular activities or may be used in adjusting glucose level of blood. Hence, glycogenolysis helps the body to regulate blood glucose level.

2.7 ELECTRON TRANSPORT SYSTEM (ETS) OVERVIEW

It is the last step in aerobic respiration. It occurs on the cristae and F_1 particles of mitochondria. During respiration NAD and FAD accepts hydrogen and get reduced to $NADH_2$ and $FADH_2$. These are finally oxidized by free molecular oxygen to produce ATP. It requires respiratory chain (ETC). The ETS consist of various enzyme and co-enzymes which acts as hydrogen and electron carriers. All co-enzymes and cytochromes are arranged in orderly manner in specific sequence according to their increasing redox potential (i.e. the ability to give or take up electrons). The electron acceptors in ETS are arranged in order of decreasing energy level in F_1 particles. NAD is the first and oxygen is the last component in ETS. It involves seven electron carriers sequentially located in the inner membrane of mitochondria. NAD and FMN as coenzyme, coenzyme Q or Ubiquinone, cytb, $cytc_1$, cytc, cyta, $cyta_3$. The reduced coenzymes pass the electrons in ETS by two routes.

In the first route, $NADH_2$ passes electrons to FMN and then to coenzyme Q. In the second route, $FADH_2$ passes electrons directly to coenzymes. Q. Co-Q passes these electron to cytochromes arranged in sequence cytb, c_1, a and a_3 and then to O_2. FMN and FAD are the first electron carriers in route I and II respectively. Thus, after oxidation, $NADH_2$ produces 3 ATP and $FADH_2$ produces 2 ATP molecules.

Hydrogen and Electron Transport through ETS :

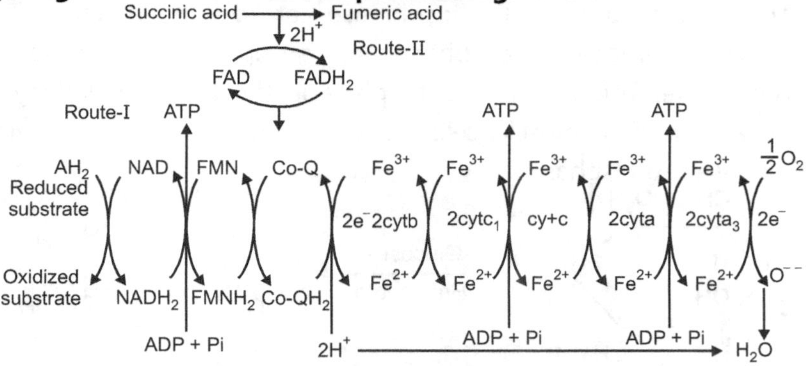

Fig. 2.9 : Electron Transport System (ETS)

In ETS, hydrogen atoms are accepted by NAD which get reduced to $NADH_2$ (from unknown hydrogen acceptor A). The hydrogen then released by $NADH_2$ on oxidation are passed to FMN which get reduced to $FMNH_2$. During this oxidation-reduction process, energy is released which is used in the formation of ATP from ADP + Pi. The $FMNH_2$ is oxidized to FMN by removal of hydrogen. Here, hydrogen undergo ionization or splits into two protons ($2H^+$) and two electrons ($2e^-$).

$$H_2 \rightarrow 2H^+ + 2e^-$$

The protons remains in mitochondrial matrix and electrons ($2e^-$) are transferred to cO-Q or ubiquinone. The CO-Q takes these two electrons and 2 protons ($2H^+$) from the matrix and becomes $CO\text{-}QH_2$. Reduced ubiquinone transfer its electron to cytochrome b and $2H^+$ to mitochondrial matrix.

The $FADH_2$ formed in Kreb's cycle also enters into electrons transport chain at this stage by transferring its $2H^+$ to CO-Q and CO-Q is reduced to $cO\text{-}QH_2$. The Reduced $CO\text{-}QH_2$ transfer its electrons to cytochrome system and $2H^+$ to matrix.

Cytochrome a and a_3 (cyta-a_3) are together called as cytochrome oxidase complex. Cyta and $Cyta_3$ has additional two copper center which helps in transfer of electron to oxygen. In cytochrome system, when the two electrons transferred, Fe^{+++} and F^{++} ions of cytochrome undergo reduction and oxidation respectively.

$$Fe^{+++} + e^- \rightarrow Fe^{++} \text{ (ferrous state) Reduction}$$

$$Fe^{++} - e^- \rightarrow Fe^{+++} \text{ (ferric state) Oxidation}$$

During transfer of electrons, energy is released, which is utilized in the formation of 2 ATP molecules from ADP + iP. Second ATP is formed between $Cytc_1$ and Cytc and third ATP is formed between Cyta and $Cyta_3$.

Finally the electrons from cytrochrome system are accepted by atmospheric molecular oxygen which get ionized to (O^{--}) i.e. $\frac{1}{2}$, O_2^{--} .

Thus oxygen is the last electron acceptor in ETS. The ionized oxygen $\left(\frac{1}{2}O_2^{--}\right)$ combines with $2H^+$ ions from the matrix to form water. This water is called metabolic water.

$$\frac{1}{2}, O_2^{--} + 2H^+ \rightarrow H_2O$$

The process of oxidation of ionized oxygen (O^{--}) and $2H^+$ to form water molecule is called terminal oxidation. The formation of ATP is termed as oxidative phosphorylation because the energy used in ATP synthesis is obtained by oxidation of reduced co-enzymes. For one $NADH_2$, 3 ATPs and for one $FADH_2$, 2 ATPs are generated on ETS. Thus, total 34 ATPs are synthesize during ETS from hydrogens generated from one glucose molecule in respiration.

Significance of ETS :

1. In ETS, glucose is completely oxidized in stepwise manner.
2. Through oxidative phosphorylation about 34 ATP molecules are generated in ETS from one glucose molecules.
3. Damage of cell is prevented due to stepwise and controlled release of energy.
4. It helps in recyclining of NAD and FAD in cell.
5. It generates metabolic water molecule at the end which is provided to Kreb's cycle for its operation.

EXERCISE

1. Define glycolysis and explain in detail reactions involved in glycolysis.
2. Describe Kreb's cycle.
3. Explain in brief Gluconeogenesis of Amino Acids and Propionic Acid
4. Describe Hydrogen and Electron Transport through ETS.
5. Write short notes on the following:
   (a) Phosphorylation in glycolysis
   (b) Oxidative Phosphorylation
   (c) ATP generation

(d) Significance of Glycolysis

(e) Significance of Kreb's Cycle

(f) Glycogenolysis

(g) Gluconeogenesis

(h) Significance of Gluconeogenesis

(i) Significance of ETS

3

LIPID METABOLISM

3.1 INTRODUCTION

Biological lipids are a chemically diverse group of compounds. They are insoluble in water which is common and defining feature of lipids. They are diverse in chemical nature as well as their biological functions. Fats and oils are the principal forms of lipids which store energy in many organisms. Phospholipids and sterols are major structural lipids of biological membranes. Other lipids, although present in relatively small quantities, play crucial roles as enzyme cofactors, electron carriers, light absorbing pigments, hydrophobic anchors for proteins, "chaperones" to help membrane proteins fold, emulsifying agents in the digestive tract, hormones and intracellular messengers.

3.2 BIOSYNTHESIS OF FATTY ACID

The ability to synthesize a variety of lipids is essential to all organisms. Like other biosynthetic pathways, these reaction sequences are endergonic and reductive. They use ATP as a source of metabolic energy and a reduced electron carrier (usually NADPH) as reductant. The fatty acid biosynthesis and breakdown of fatty acids occur by different pathways, hence they are catalyzed by different sets of enzymes, and also they take place in different parts of the cell. The biosynthesis of fatty acids requires enzyme Fatty acid synthase complex and a three-carbon intermediate, malonyl-Co A.

3.2.1 Enzyme Fatty Acid Synthase Complex

The synthesis of fatty acids is carried out by the enzyme Fatty acid synthase. This enzyme is having seven active sites for seven different functions of fatty acid synthesis. Its formation is also different in different organisms (Table 3.1, Fig. 3.1).

In *E. coli* and some plants, the enzyme is formed of seven separate polypeptides while the fatty acid synthases of yeast is formed of two large, multifunctional polypeptides, with three activities on the α subunit and four on the β subunit. The vertebrate enzyme is also multienzyme complex formed of a single large polypeptide (*MW* - 240,000). It contains all seven enzymatic activities as well as a hydrolytic activity that cleaves the finished fatty acid from the ACP-like part of the enzyme complex.

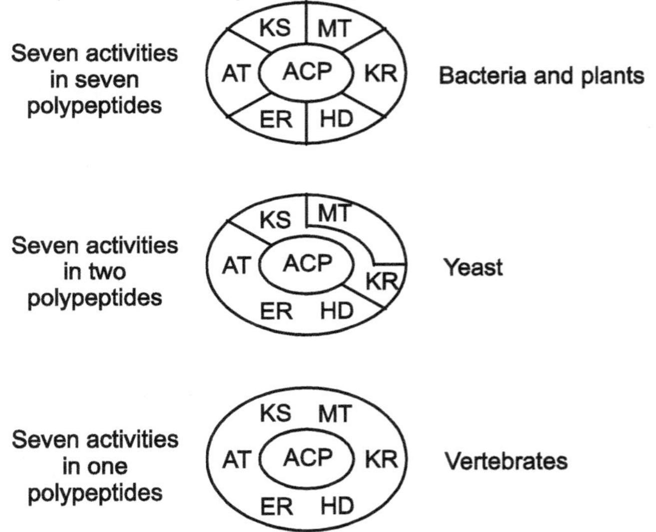

Fig. 3.1 : Structure of fatty acid synthase enzyme

Table 3.1 : Comparison of Fatty acid synthase of different organisms

Sr. No.	Organism	Number of polypeptides	Number of active sites
1	*E. coli* and Plants	07 (3 others act at some stage of the process)	07
2	Yeast	02 (α - 3 active sites β - 4 active sites	07
3	Vertebrates	01	07 and Hydrolytic activity at the end of process

The core of the *E. coli* fatty acid synthase system consists of seven separate polypeptides (Table 3.2), and at least three others act at some stage of the process. The proteins act together to catalyze the formation of fatty acids from acetyl-Co A and malonyl-Co A. Throughout the process, the intermediates products remain covalently attached to the synthase complex.

Table 3.2 : Proteins of the Fatty Acid Synthase Complex of *E. coli*

Sr. No.	Component	Symbol	Function
1	Acyl carrier protein	ACP	Carries acyl groups in thioester linkage
2	Acetyl-Co A–ACP transacetylase	AT	Transfers acyl group from Co A to Cysteine residue of KS
3	β Ketoacyl-ACP synthase	KS	Condenses acyl and malonyl groups (KS has at least three isozymes)
4	Malonyl-Co A–ACP transferase	MT	Transfers malonyl group from Co A to ACP
5	β Ketoacyl-ACP reductase	KR	Reduces β keto group to β hydroxyl group
6	β Hydroxyacyl-ACP dehydratase	HD	Removes H_2O from β hydroxyacyl-ACP, creating double bond
7	Enoyl-ACP reductase	ER	Reduces double bond, forming saturated acyl-ACP

3.2.2 Synthesis of Malonyl-Co A

The malonyl-Co A is formed from acetyl-Co A and carboxyl group derived from bicarbonate in all types of cells. This is two step irreversible process, catalyzed by an enzyme acetyl-Co A carboxylase. In animal cells, this enzyme has single multifunctional polypeptide containing a biotin prosthetic group. The enzyme has three functional groups as Biotin carrier protein, Biotin carboxylase and Transcarboxylase. The biotin is covalently bound to the Lysine amino acid of Biotin carrier protein.

In first step the carboxyl group, derived from bicarbonate (HCO_3^-) is first transferred to biotin in an ATP dependent reaction. The biotin group serves as a temporary carrier of CO_2. Then in the second step the enzyme transfers the CO_2 to acetyl-Co A to yield malonyl-Co A.

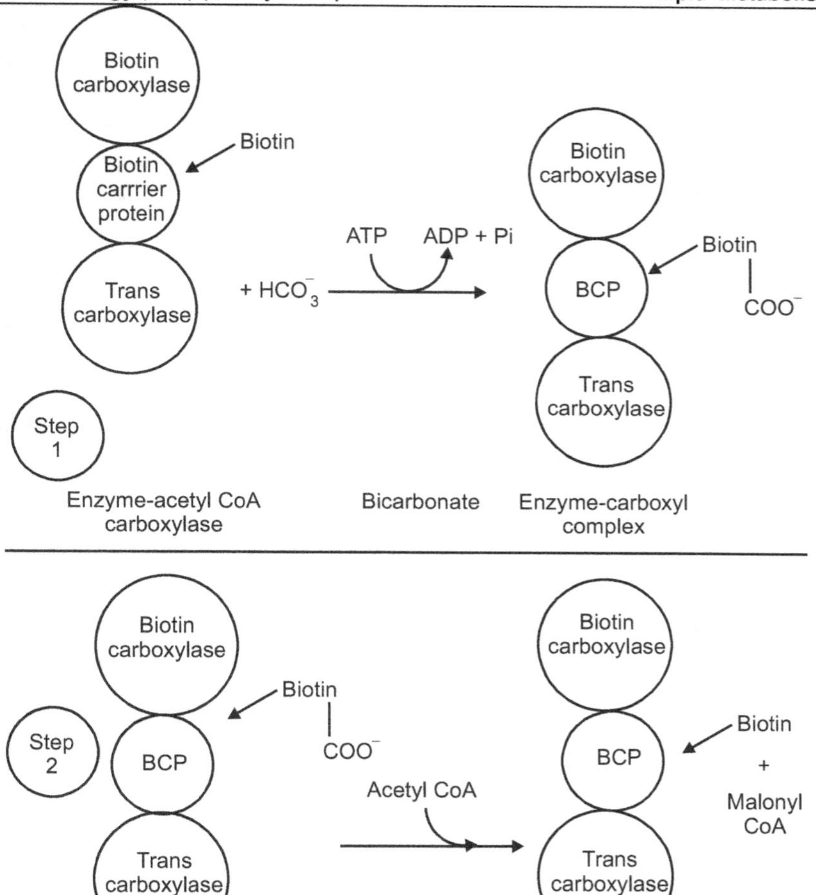

Fig. 3.2 : Two step reaction of synthesis of malonyl CoA from Acetyl Co-A by addition of carboxyl group

Fig. 3.3 : Malonyl CoA

3.2.3 Synthesis of Fatty Acids

The synthesis of fatty acids is a repetitive process of addition of Malonyl group from Malonyl Co A on the Fatty acid synthase enzyme. This is the condensation reaction which results in joining of different malonyl groups in a chain. The process begins with the charging of two thiol groups of the enzyme by the two acyl groups - acetyl and malonyl. First, the acetyl group of acetyl- Co A is transferred to the β ketoacyl-ACP synthase (KS). This reaction is catalyzed by acetyl-Co A ACP transacetylase (AT). The second reaction is the transfer of the malonyl group from malonyl-Co A to the Acyl Carrier Protein (ACP). It is catalyzed by malonyl-Co A ACP transferase (MT). In above both reactions the acetyl and malonyl groups attaches to the - SH of the respective enzymes of complex. In the charged synthase complex, the acetyl and malonyl groups are very close to each other and are activated for the chain-lengthening process. Now the enzyme has acetyl and malonyl groups on the β ketoacyl-ACP synthase (KS) and Acyl Carrier Protein (ACP) respectively. The formation of bond between these two groups takes place in five steps as described below. Further elongation of the chain is repeated by the addition of malonyl group to ACP and formation of bond.

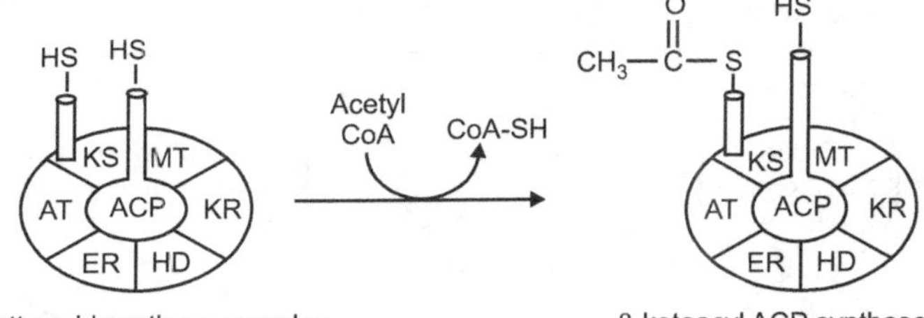

Fig. 3.4 : Charging of fatty acyl synthase enzyme by acyl and malonyl groups

3.2.3.1 Step 1 Condensation

The first reaction in the formation of a fatty acid chain is condensation of the activated acetyl and malonyl groups. CO_2 from malonyl is removed and the acetyl group is transferred from the β ketoacyl-ACP synthase (KS) enzyme to the malonyl group on the ACP. This transfer results in the formation of acetoacetyl-ACP which has two carbon units at the methyl-terminal and this reaction is catalyzed by β ketoacyl-ACP synthase (KS).

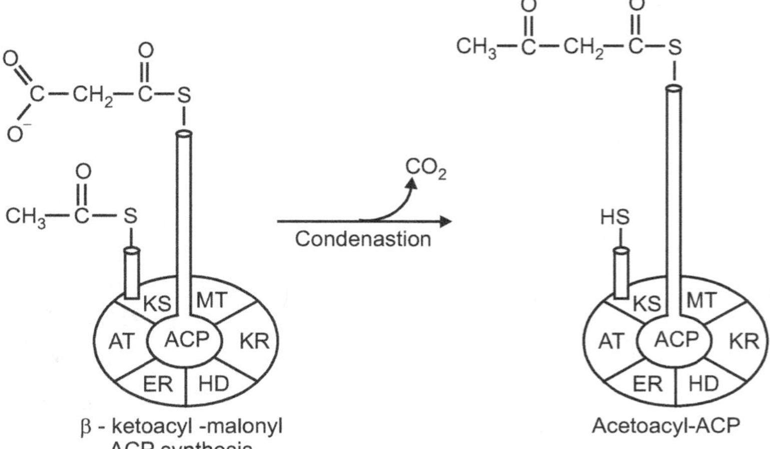

Fig. 3.5 : Step 1 condensation

3.2.3.2 Step 2 Reduction of the Carbonyl Group

The acetoacetyl ACP formed in the condensation step undergoes reduction of the carbonyl group at C-3. This reaction is brought about by β ketoacyl-ACP reductase (KR). For this reduction two electrons are donated by $NADPH^+ + H^+$ which is converted in to NADP. The final reduced product is D β hydroxybutyryl-ACP.

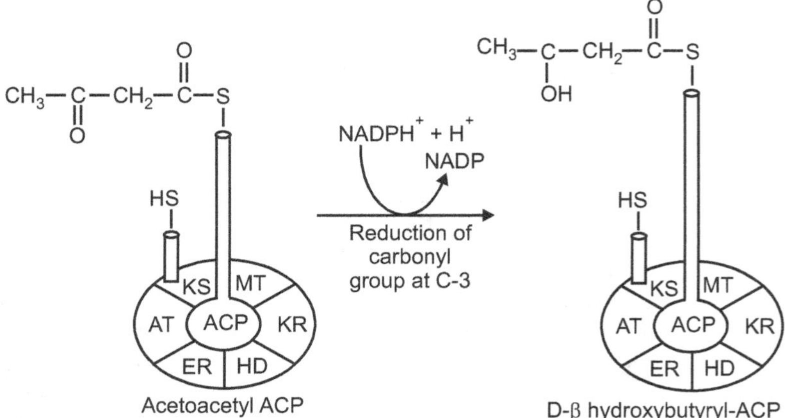

Fig. 3.6 : Step 2 Reduction

3.2.3.3 Step 3 Dehydration

One molecule of water is removed by the union of H from C-2 and OH from C-3 of D β hydroxybutyryl-ACP. A double bond is formed between these two carbons to form *trans* Δ^2 butenoyl ACP. This dehydration reaction is catalyzed by β hydroxyacyl-ACP dehydratase (HD).

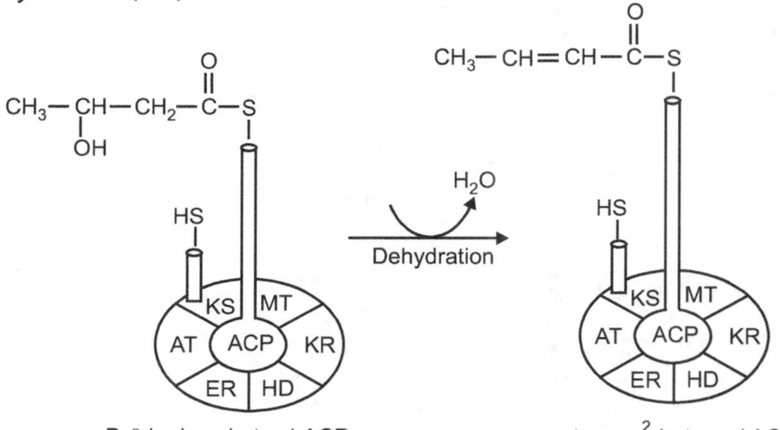

Fig. 3.7 : Step 3 Dehydration

3.2.3.4 Step 4 Reduction of the Double Bond

In the fourth step, the double bond of *trans* Δ^2 butenoyl ACP is reduced (saturated) by receiving two electrons form NADPH$^+$ + H$^+$. The enzyme involved is enoyl-ACP reductase (ER) and the product is Butyryl-ACP.

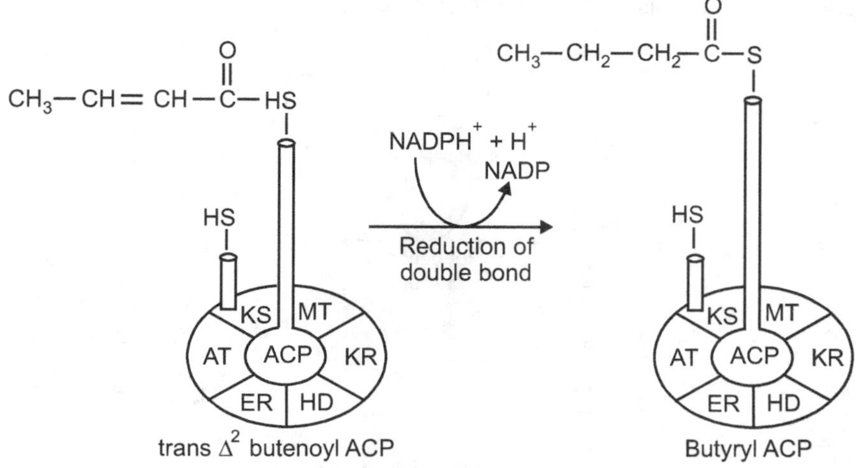

Fig. 3.8 : Step 4 Reduction of double bond

3.2.3.5 Step 5 Translocation of Butyryl ACP

The final step in the addition of two Carbons to the growing fatty acid chain is the translocation of Butyryl-ACP to the β ketoacyl-ACP synthase (KS). This reaction is catalyzed by Acetyl-Co A–ACP transacetylase (AT).

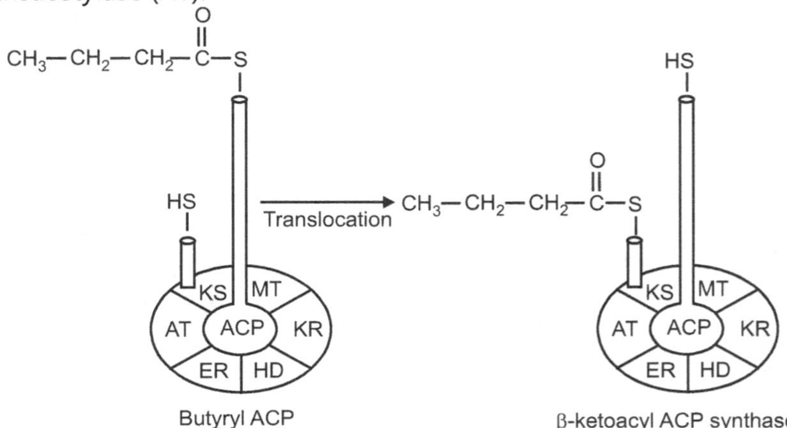

Fig. 3.9 : Step 5 Translocation

Overall five steps of fatty acid synthesis sequence results in the addition of two Carbons to the growing fatty acid chain. In all these steps one molecule of CO_2 and one molecule of H_2O is removed while two molecules of $NADPH^+ + H^+$ are used. They donate two pairs of electrons (H^+) in the overall reaction.

3.2.4 The Fatty Acid Synthase Reactions are Repeated

Production of the four-carbon, saturated fatty acyl–ACP completes one pass through the fatty acid synthase complex. To start the next cycle of five reactions that lengthens the chain by two more carbons, another malonyl group is linked to the ACP. Condensation occurs as the butyryl group, acting like the acetyl group in the first cycle, is linked to two carbons of the malonyl-ACP group with concurrent loss of CO_2. The product of this condensation is a six carbon acyl group, covalently bound to the ACP. Its β keto group is reduced in the next three steps of the synthase cycle to yield the saturated acyl group, exactly as in the first round of reactions - in this case forming the six-carbon product. Seven cycles of condensation and reduction produce the 16-carbon saturated palmitoyl group, still bound to ACP. Now the chain elongation by the synthase complex stops and free palmitate is released from the ACP by a hydrolytic activity in the complex.

The overall reaction for the synthesis of palmitate from acetyl-Co A takes place in two parts as follows.

First, the formation of seven malonyl-Co A molecules:

$$7 \text{ Acetyl Co A} + 7CO_2 + 7ATP \longrightarrow 7 \text{ malonyl Co A} + 7ADP + 7 Pi$$

then, seven cycles of condensation and reduction:

$$\text{Acetyl-CoA} + 7\text{malonylCoA} + 14NADPH^+ + 14H^+ \longrightarrow \text{Palmitate} + 7CO_2 + 8 \text{ CoA} + 14NADP^+ + 6H_2O$$

The overall process is

$$8 \text{ Acetyl Co A} + 7ATP + 14NADPH^+ + 14H^+ \longrightarrow \text{Palmitate} + 8 \text{ Co A} + 7ADP + 7Pi + 14NADP^+ + 6H_2O$$

The biosynthesis of fatty acids such as palmitate thus requires acetyl-Co A and the input of chemical energy in two forms: the group transfer potential of ATP and the reducing power of NADPH. The ATP is required to attach CO_2 to acetyl-Co A to make malonyl-Co A; the NADPH is required to reduce the double bonds.

3.3 OXIDATION OF FATTY ACIDS

The complete oxidation of fatty acids to CO_2 and H_2O provides about 9 kcal (38 kJ) of energy per gram. This process of oxidation takes place in three stages.

(i) The oxidation of long-chain fatty acids to two-carbon fragments, in the form of acetyl-Co A (β oxidation) This process takes place in Mitochondria.

(ii) The oxidation of acetyl-Co A to CO_2 in the citric acid cycle (TCA Cycle - Mitochondria).

(iii) The transfer of electrons from reduced electron carriers to the mitochondrial respiratory chain (Electron Transport System).

3.3.1 Activation of fatty acids

The free fatty acids cannot pass from cytosol (cytoplasm) to mitochondria directly through the mitochondrial membrane. Hence for their transport they must be activated. The activation of carboxyl group at C-1 of fatty acid is carried out in a series of three enzymatic reactions. Out of these one takes place in Cytosol and two in the mitochondrial membrane by carnitine shuttle.

The first enzymatic reaction takes place in the cytosol and it is catalyzed by the acyl-Co A synthatase which convert fatty acid in to fatty acyl-Co A. One molecule of ATP is used for the attachment of Co A to the fatty acid. The overall reaction takes place in two steps.

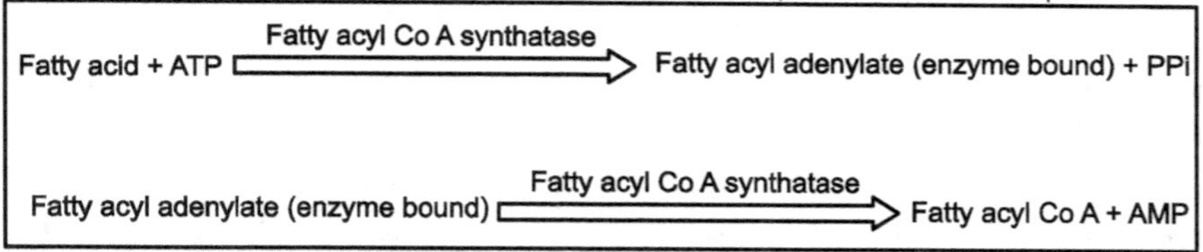

Fig. 3.10 : Activation of fatty acids

3.3.2 Transport of Fatty acyl-CoA Synthase in Mitochondria

The transfer of fatty acyl-CoA into mitochondria needs carnitine acyl-transferase I and carnitine acyl-transferase II as the fatty acyl-CoA esters do not pass the mitochondrial membranes. These two enzymes are present in the outer and inner mitochondrial membranes respectively and bring about two reactions.

The second reaction takes place on the outer mitochondrial membrane where Co A is removed from the fatty acyl-Co A by the carnitine acyl-transferase I to form fatty acyl-carnitine. The fatty acyl group transiently attaches to hydroxyl group of carnitine to form fatty acyl-carnitine. In the third and final reaction of the carnitine shuttle, the fatty acyl group is enzymatically transferred from carnitine to intramitochondrial Co A by carnitine acyl-transferase II. This isozyme, located on the inner face of the inner mitochondrial membrane, regenerates fatty acyl–Co A and releases it, along with free carnitine, into the matrix. Carnitine reenters the intermembrane space via the acyl-carnitine/carnitine transporter.

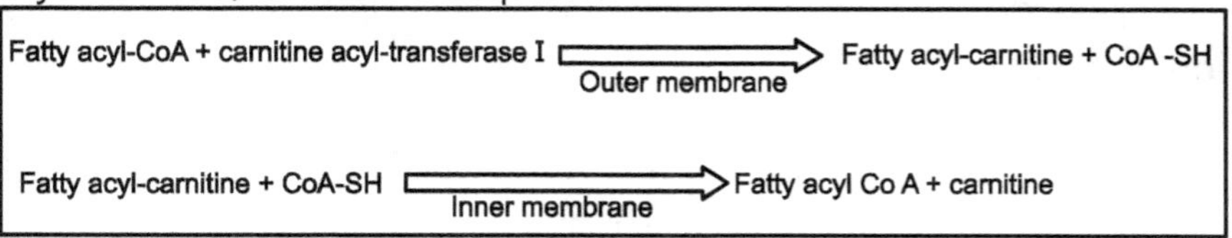

Fig. 3.11 : Transport of fatty acids in to mitochondria

The fatty acyl–Co A formed is now active and ready to enter in the β oxidation.

3.3.3 β oxidation

The process of β oxidation takes place in the mitochondrial matrix. The C-C single bond in the fatty acids is relatively stable than the double bonds. To overcome this stability of the C-C bonds in a fatty acid, the carboxyl group at C-1 is activated by attachment to coenzyme A. This attachment allows stepwise oxidation of the fatty acyl group at the C-3 or β position. Hence, this process is known as β oxidation (it is a four step process).

The β oxidation of fatty acids such as 16-Carbon Palmitic acid undergoes oxidative removal of successive two-carbon units in the form of acetyl-Co A. The process starts from the carboxyl end of the fatty acyl chain. The 16-carbon palmitic acid (palmitate at pH 7) undergoes seven passes through the oxidative sequence, in each pass losing two carbons as acetyl-Co A. At the end of seven cycles the last two carbons of palmitate (originally C-15 and C-16) remain as acetyl-Co A. The overall result is the conversion of the 16-carbon chain of palmitate to eight two-carbon acetyl groups of acetyl-Co A molecules (Fig. 3.12). Formation of each acetyl-Co A requires removal of four hydrogen atoms (two pairs of electrons and four H) from the fatty acyl moiety by dehydrogenases.

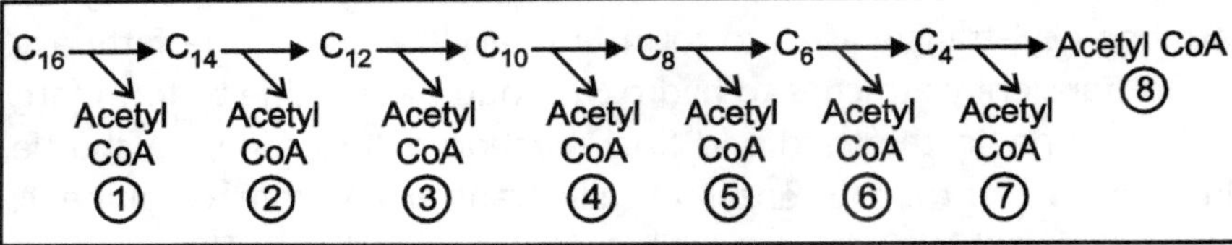

Fig. 3.12 : β oxidation of 16-C Palmitic acid to produce eight molecules of Acetyl Co A

Overall process of β oxidation takes place in four steps.

3.3.3.1 Step 1 Dehydrogenation

In the first step, dehydrogenation of fatty acyl–Co A produces a double bond between α and β carbon atoms (C-2 and C-3), yielding a *trans*-Δ^2-enoyl-Co A (the symbol Δ^2 designates the position of the double bond). Two H^+ ions are removed from α and β carbon atoms which are received by FAD. Thus FAD is reduced to $FADH_2$. This reduced form of the dehydrogenase immediately donates its electrons to an electron carrier of the mitochondrial respiratory chain, the electron-transferring flavoprotein (ETF). FAD is the electron acceptor, and electrons from the reaction ultimately enter the ETS. The new double bond has the trans configuration, whereas the double bonds in naturally occurring unsaturated fatty acids are normally in the cis configuration. This first step is catalyzed by three isozymes of acyl-Co A dehydrogenase, each specific for a range of fatty-acyl chain lengths: very-long-chain acyl-Co A dehydrogenase (VLCAD), acting on fatty acids of 12 to 18 carbons; medium-chain (MCAD), acting on fatty acids of 4 to 14 carbons; and short-chain (SCAD), acting on fatty acids of 4 to 8 carbons.

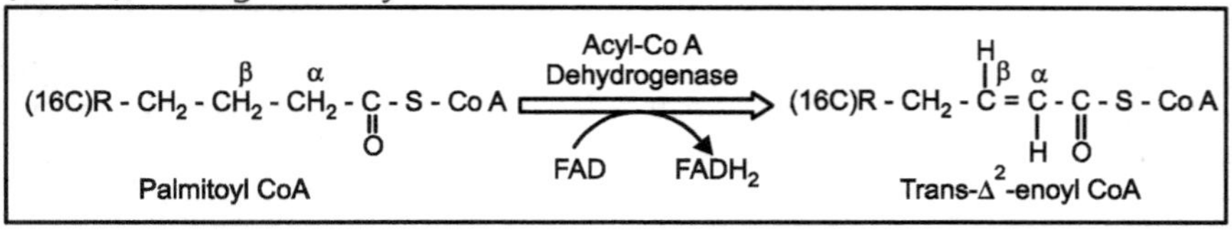

Fig. 3.13 : Dehydrogenation of fatty acyl–Co A

3.3.3.2 Step 2 Hydration

In the second step of the β oxidation cycle, water is added to the double bond of the *trans*-Δ^2-enoyl-Co A to form the L stereoisomer of β hydroxyacyl-Co A (3-hydroxyacyl-Co A). This reaction, catalyzed by enoyl-Co A hydratase, is formally analogous to the fumarase reaction in the citric acid cycle, in which H_2O adds across the α and β double bond.

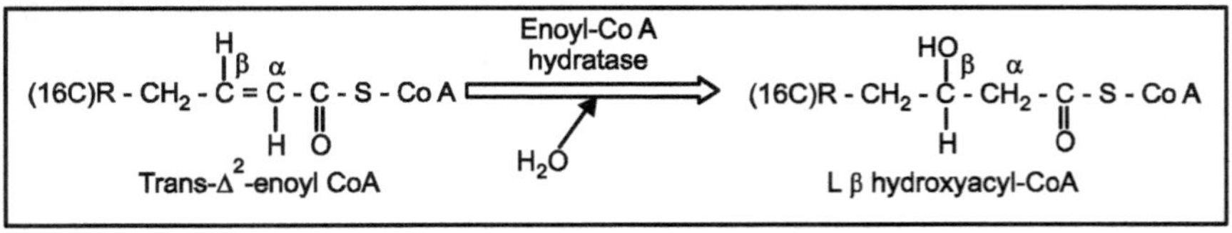

Fig. 3.14 : Hydration

3.3.3.3 Step 3 Dehydrogenation

In the third step, L β hydroxyacyl-Co A is dehydrogenated to form β ketoacyl-Co A, by the action of β hydroxyacyl-Co A dehydrogenase; NAD^+ is the electron acceptor. The NADH formed in the reaction donates its electrons to NADH dehydrogenase, an electron carrier of the respiratory chain, and ATP is formed from ADP as the electrons pass to O_2. The reaction catalyzed by β hydroxyacyl-Co A dehydrogenase is closely analogous to the malate dehydrogenase reaction of the citric acid cycle.

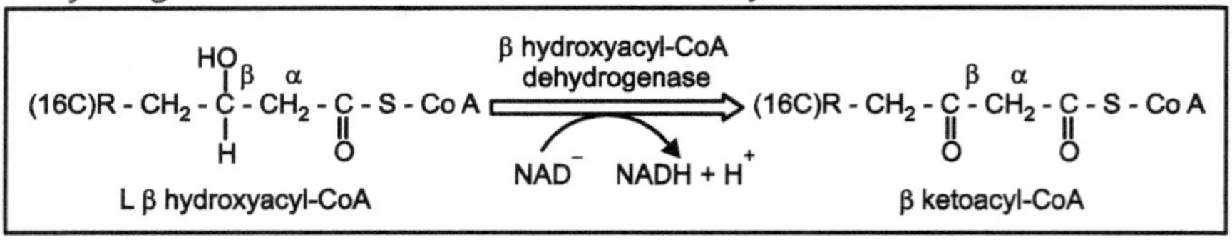

Fig. 3.15 : Dehydrogenation

3.3.3.4 Step 4 Thiolysis

The fourth and last step of the β oxidation cycle is catalyzed by acyl-Co A acetyltransferase, more commonly called thiolase, which promotes reaction of β ketoacyl-Co A with a molecule of free coenzyme A to split off the carboxyl-terminal two-carbon fragment of the original fatty acid as acetyl-Co A. The other product is the coenzyme A thioester of the fatty acid, now shortened by two carbon atoms. This reaction is called thiolysis, by analogy with the process of hydrolysis, because the β ketoacyl-Co A is cleaved by reaction with the thiol group of coenzyme A.

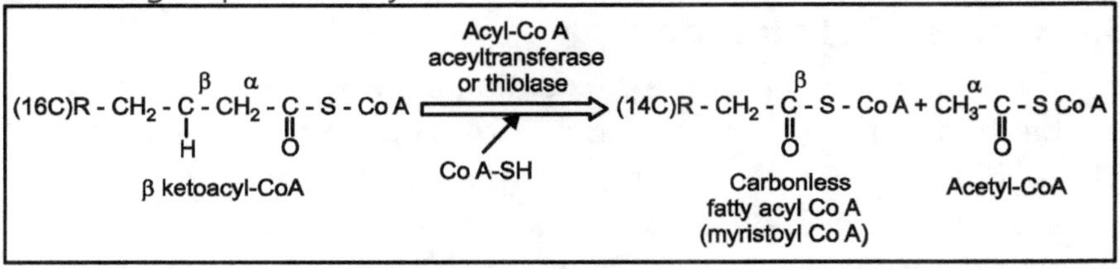

$$(16C)R - CH_2 - \overset{\beta}{\underset{H}{C}} - \overset{\alpha}{CH_2} - \underset{O}{C} - S - CoA \xrightarrow[\text{Co A-SH}]{\substack{\text{Acyl-Co A} \\ \text{aceyltransferase} \\ \text{or thiolase}}} (14C)R - CH_2 - \overset{\beta}{\underset{O}{C}} - S - CoA + \overset{\alpha}{CH_3} - \underset{O}{C} - S\ CoA$$

β ketoacyl-CoA $\qquad\qquad\qquad$ Carbonless fatty acyl Co A (myristoyl Co A) $\qquad$ Acetyl-CoA

Fig. 3.16 : Thiolysis of fatty acyl–Co A

Overall four steps of β oxidation sequence results in formation of one molecule of acetyl-Co A. Other products are two pairs of electrons, and four protons (H^+). The fatty acyl Co A is shortened by two carbon atoms. The overall equation for one pass, beginning with the coenzyme A ester of a fatty acid is as follows.

Fatty acyl-Co A + Co A + FAD + NAD^+ + H_2O $\longrightarrow$ 2 Carbon less Fatty acyl Co A + acetyl-Co A + $FADH_2$ + NADH + H^+

3.3.4 The Four β Oxidation Steps are Repeated to Yield Acetyl-Co A and ATP

After the removal of one acetyl-Co A unit from palmitoyl- Co A, the coenzyme A thioester of the shortened fatty acid (now the 14-carbon myristate) remains. The myristoyl-Co A can now go through another set of four β oxidation reactions, exactly analogous to the first, to yield a second molecule of acetyl-Co A and lauroyl-Co A, the coenzyme A thioester of the 12-carbon laurate. Altogether, seven passes through the β oxidation sequence are required to oxidize one molecule of palmitoyl-Co A to eight molecules of acetyl-Co A. The overall equation is

Palmitoyl-Co A + 7Co A + 7FAD + $7NAD^+$ + $7H_2O$ $\longrightarrow$ 8 acetyl-Co A + $7FADH_2$ + 7NADH + $7H^+$

Each molecule of $FADH_2$ formed during oxidation of the fatty acid donates a pair of electrons to ETF of the respiratory chain, and about 1.5 molecules of ATP are generated during the ensuing transfer of each electron pair to O_2. Similarly, each molecule of NADH formed delivers a pair of electrons to the mitochondrial NADH dehydrogenase, and the subsequent transfer of each pair of electrons to O_2 results in formation of about 2.5 molecules of ATP. Thus four molecules of ATP are formed for each two-carbon unit removed in one pass through the sequence. Note that water is also produced in this process. Transfer of electrons from NADH or $FADH_2$ to O_2 yields one H_2O per electron pair. Reduction of O_2 by NADH also consumes one H^+ per NADH molecule:

$$NADH + H^+ + 1/2\ O_2 \longrightarrow NAD^+ + H_2O.$$

In hibernating animals, fatty acid oxidation provides metabolic energy, heat, and water. These all are essential for survival of an animal that neither eats nor drinks for long periods. Camels obtain water to supplement the meager supply available in their natural environment by oxidation of fats stored in their hump. The overall equation for the oxidation of palmitoyl- Co A to eight molecules of acetyl-Co A, including the electron transfers and oxidative phosphorylation, is

$$\text{Palmitoyl-Co A} + 7\text{Co A} + 7O_2 + 28Pi + 28ADP \longrightarrow 8\text{ acetyl-Co A} + 28ATP + 7H_2O$$

EXERCISE

1. Explain biosynthesis of fatty acid.
2. Compare fatty acid synthase of yeast and E. coli.
3. Write the reaction steps of synthesis of malonyl
4. Write in detail on OXIDATION OF FATTY ACIDS
5. Write short notes on the following:
   (a) Enzyme Fatty Acid Synthase Complex
   (b) Synthesis of Malonyl-Co A
   (c) Synthesis of Fatty Acids
   (d) Activation of fatty acids
   (e) β oxidation

☆☆☆

4

CHAPTER

PROTEIN METABOLISM

4.1 INTRODUCTION

Proteins are the most abundant biological macromolecules, occurring in all cells. They also occur in great variety and of different kinds, ranging in size from relatively small peptides to huge polymers with molecular weights in the millions. Moreover, proteins exhibit enormous diversity of biological function and are the most important final products of the information pathways. Proteins are the molecular instruments through which genetic information is expressed.

All proteins are made up from the set of 20 amino acids called as building blocks which are covalently linked in characteristic linear sequences. Each of these 20 amino acids has a side chain with distinctive chemical properties. Hence this group of 20 precursor molecules may be regarded as the alphabet in which the language of protein structure is written. The cells can produce proteins with strikingly different properties and activities by joining the same 20 amino acids in many different combinations and sequences. The diverse products or components of the cells and organisms such as enzymes, hormones, antibodies, transporters, muscle fibers, the lens protein of the eye, feathers, spider webs, rhinoceros horn, milk proteins, antibiotics, mushroom poisons, and number of others are made up of these building blocks. Among these protein products, the enzymes are the most varied and specialized. Virtually all cellular reactions are catalyzed by enzymes.

4.2 AMINO ACIDS

The Amino acids have a carboxyl group and an amino group bonded to the single carbon atom. This carbon is called as α carbon. All the amino acids differ from each other in their side chains, or R groups, which vary in structure, size, and electric charge. The R groups influence the solubility of the amino acids in water. The common amino acids of proteins have been assigned three-letter abbreviations and one-letter symbols (Table 1). These abbreviations and symbols are used as shorthand to indicate the composition and sequence of amino acids polymerized in proteins.

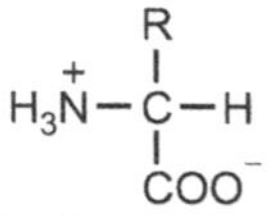

$$\overset{\displaystyle R}{\underset{\displaystyle COO^-}{\overset{+}{H_3N}-C-H}}$$

Fig. 4.1 : General structure of an Amino acid

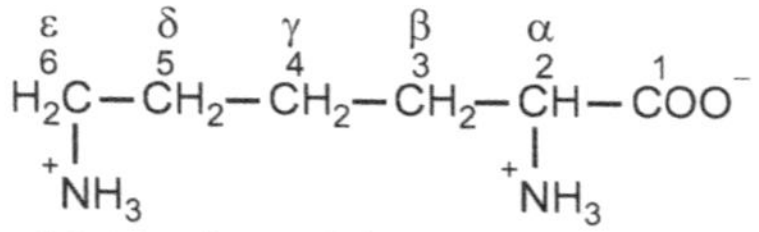

$$\overset{\varepsilon}{\underset{6}{H_2C}}-\overset{\delta}{\underset{5}{CH_2}}-\overset{\gamma}{\underset{4}{CH_2}}-\overset{\beta}{\underset{3}{CH_2}}-\overset{\alpha}{\underset{2}{CH}}-\overset{1}{COO^-}$$

Fig. 4.2 : Amino acid - Lysine showing two methods of nomenclature of Carbon

All the amino acids are grouped into five main classes based on the properties of their R groups (Table 1). They are classified mainly on their polarity or tendency to interact with water at biological pH (near pH 7.0). The polarity of the R groups varies widely, from nonpolar and hydrophobic (water-insoluble) to highly polar and hydrophilic (water-soluble). Within each class there are gradations of polarity, size, and shape of the R groups.

All 20 of the common amino acids are α amino acids. The carbon to which amino and carboxyl group is attached is called as α carbon. The additional carbons in an R group are commonly designated β, γ, δ, ε and so forth, proceeding out from the α carbon. For most other organic molecules, carbon atoms are simply numbered from one end as C1, C2, C3, etc. Thus, by considering the numbering system, the carboxyl carbon of an amino acid would be C-1 and the α carbon would be C-2. In some cases, such as amino acids with heterocyclic R groups, the Greek lettering system is ambiguous and the numbering convention is therefore used.

Table 4.1 : Amino acids: Abbreviation, Symbol and Classification

Sr. No.	Amino Acid	Abbreviation	Symbol	Nature of R group
1	Glycine	Gly	G	Nonpolar, aliphatic
2	Alanine	Ala	A	Nonpolar, aliphatic
3	Proline	Pro	P	Nonpolar, aliphatic
4	Valine	Val	V	Nonpolar, aliphatic
5	Leucine	Leu	L	Nonpolar, aliphatic
6	Isoleucine	Ile	I	Nonpolar, aliphatic
7	Methionine	Met	M	Nonpolar, aliphatic

Conti...

8	Phenylalanine	Phe	F	Aromatic
9	Tyrosine	Tyr	Y	Aromatic
10	Tryptophan	Trp	W	Aromatic
11	Serine	Ser	S	Polar, uncharged
12	Threonine	Thr	T	Polar, uncharged
13	Cysteine	Cys	C	Polar, uncharged
14	Asparagine	Asn	N	Polar, uncharged
15	Glutamine	Gln	Q	Polar, uncharged
16	Lysine	Lys	K	Positively charged
17	Histidine	His	H	Positively charged
18	Arginine	Arg	R	Positively charged
19	Aspartate	Asp	D	Negatively charged
20	Glutamate	Glu	E	Negatively charged

4.3 TRANSAMINATION

The transfer of α amino group of most L-amino acids to α carbon atom of α ketoglutarate is called as Transamination. This reaction is catalyzed by the enzymes called aminotransferases or transaminases. In these Transamination reactions, corresponding α keto acid analog of the amino acid is left behind. There is no net deamination (loss of amino groups) in these reactions, because α ketoglutarate becomes aminated as α amino acid is deaminated. The effect of transamination reactions is to collect the amino groups from many different amino acids in the form of L-glutamate. The glutamate then functions as an amino group donor for biosynthetic pathways or for excretion pathways that lead to the elimination of nitrogenous waste products.

α-ketoglutarate + L - Amino acid ⟷ L - Glutamate + α-keto acid

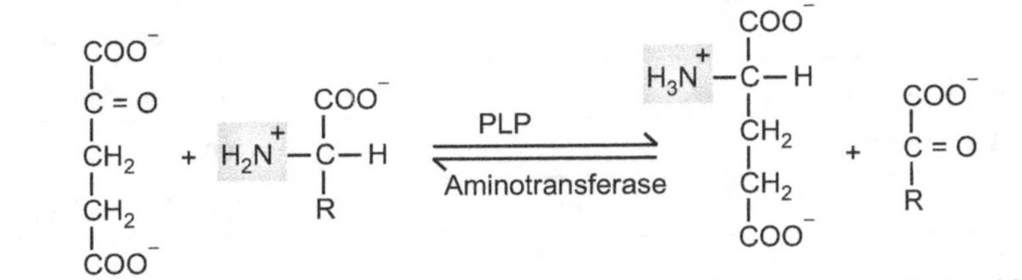

α-ketoglutarate L-amino acid L-Glutamate α-Ketoacid

Fig. 4.3 : Transamination

Cells contain different types of aminotransferases. Many are specific for α ketoglutarate as the amino group acceptor but differ in their specificity for the L-amino acid. The enzymes are named for the amino group donor (as example alanine aminotransferase, aspartate aminotransferase). The reactions catalyzed by aminotransferases are freely reversible.

All aminotransferases have the same prosthetic group and the same reaction mechanism. The prosthetic group is pyridoxal phosphate (PLP), the coenzyme form of pyridoxine, or vitamin B6. The primary role of these enzymes in cells is in the metabolism of molecules with amino groups. Pyridoxal phosphate functions as an intermediate carrier of amino groups at the active site of aminotransferases. It undergoes reversible transformations between its aldehyde form, pyridoxal phosphate, which can accept an amino group, and its aminated form, pyridoxamine phosphate, which can donate its amino group to α keto acid. Pyridoxal phosphate is generally covalently bound to the enzyme's active site through an aldimine (Schiff base) linkage to α amino group of a Lysine residue. Pyridoxal phosphate participates in a variety of reactions at α, β and γ carbons (C-2 to C-4) of amino acids. Reactions at α carbon include racemizations (interconverting L- and D-amino acids) and decarboxylations, as well as transaminations.

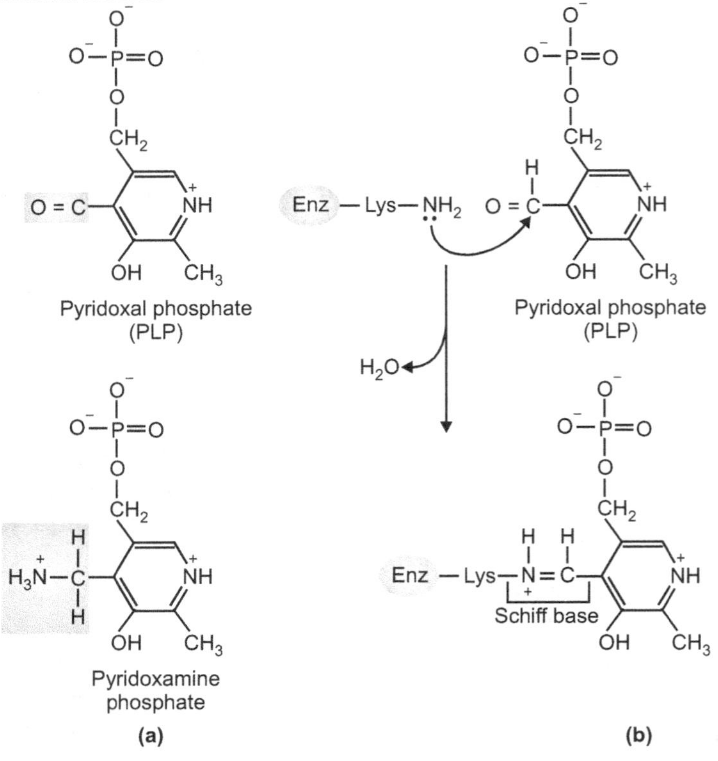

Pyridoxal phosphate
(PLP)

Pyridoxamine
phosphate

(a)

Enz — Lys — NH$_2$

Pyridoxal phosphate
(PLP)

H$_2$O

Enz — Lys — N = C

Schiff base

(b)

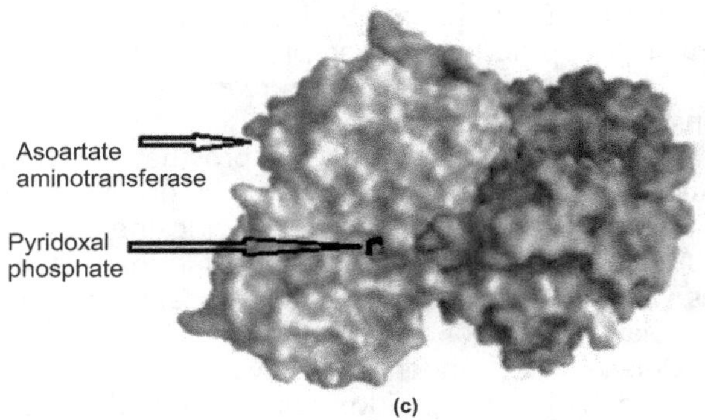

Fig. 4.4 : Pyridoxal phosphate, the prosthetic group of aminotransferases.

(a) Pyridoxal phosphate (PLP) and its aminated form, pyridoxamine phosphate, are the tightly bound coenzymes of aminotransferases. The functional groups are shaded. **(b)** Pyridoxal phosphate is bound to the enzyme through non-covalent interactions and a Schiff base linkage to a Lys residue at the active site. **(c)** PLP bound to one of the two active sites of the dimeric enzyme aspartate aminotransferase, a typical aminotransferase.

Aminotransferases (Fig. 4.4) are classic examples of enzymes catalyzing bimolecular Ping-Pong reactions, in which the first substrate reacts and the product must leave the active site before the second substrate can bind. Thus the incoming amino acid binds to the active site, donates its amino group to pyridoxal phosphate, and departs in the form of an α keto acid. The incoming α keto acid then binds, accepts the amino group from pyridoxamine phosphate, and departs in the form of an amino acid.

4.4 DEAMINATION

Deamination is the removal of an amino group from a molecule. Enzymes which catalyze this reaction are called as deaminases. In human body deamination takes place primarily in the liver. The amino group is removed from the amino acid and converted to ammonia.

There are four types of deaminations as oxidative, reductive, hydrolytic and intramolecular.

Oxidative deamination : The L-glutamate molecules formed in the transamination reactions from many of α amino acids are collected in the liver. These amino groups must be removed from glutamate to prepare them for excretion. In liver/hepatic cells (hepatocytes), glutamate is transported from the cytosol into mitochondria, where it undergoes oxidative deamination catalyzed by L-glutamate dehydrogenase. In mammals, this enzyme is present in the mitochondrial matrix. It is the only enzyme that can use either NAD+ or NADP+ as the acceptor of reducing equivalents (Fig. 18–7). The combined action of an aminotransferase and glutamate dehydrogenase is referred to as transdeamination. A few amino acids bypass the transdeamination pathway and undergo direct oxidative deamination.

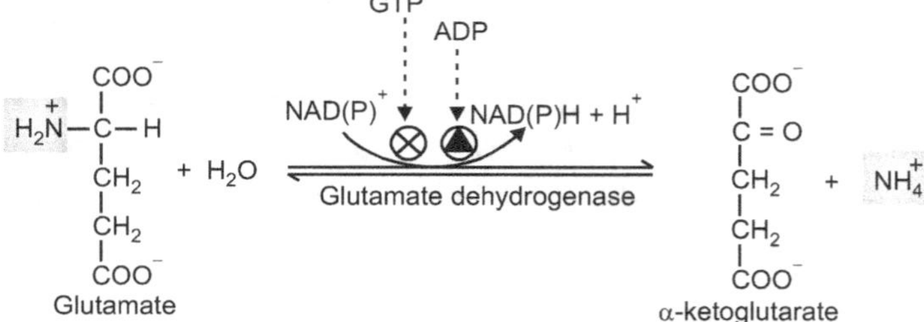

Fig. 4.5 : Oxidative deamination

The α ketoglutarate formed from glutamate deamination can be used in the citric acid cycle and for glucose synthesis. Glutamate dehydrogenase operates at an important intersection of carbon and nitrogen metabolism. It is an allosteric enzyme with six identical subunits. Its activity is influenced by a complicated array of allosteric modulators ADP and GTP. The ADP is the positive modulator and GTP is the negative modulator. Mutations that alter the allosteric binding site for GTP or cause permanent activation of glutamate dehydrogenase lead to a human genetic disorder called hyperinsulinism-hyperammonemia syndrome, characterized by elevated levels of ammonia in the bloodstream and hypoglycemia.

Reductive deamination :

$$R\text{-}CH(NH_2)\text{-}COOH + 2H^+ \longrightarrow R\text{-}CH_2\text{-}COOH + NH_3$$

          Amino acid   Fatty acid   Ammonia

Hydrolytic deamination :

$$R\text{-}CH(NH_2)\text{-}COOH + H_2O \longrightarrow R\text{-}CH(OH)\text{-}COOH + NH_3$$

          Amino acid   Hydroxy acid   Ammonia

Intramolecular deamination :

$$R - CH(NH_2) - COOH \longrightarrow R - CH = CH - COOH \;+\; NH_3$$

Amino acid   Unsaturated fatty acid   Ammonia

4.5 UREA CYCLE

The amino groups from amino acids released by the deamination reactions are the ammonia molecules which are excreted out from the body in various forms as ammonia, urea, uric acid, etc by different animals. In most terrestrial animals including humans the ammonia is converted in to urea in the liver cells. They are called as ureotelic animals. The ammonia deposited in the mitochondria of hepatocytes is converted to urea. The process takes place in a cyclic manner called as urea cycle. This pathway was discovered in 1932 by Hans Krebs (who later also discovered the citric acid cycle) and his medical student associate, Kurt Henseleit. Urea production occurs almost exclusively in the liver and is the fate of most of the ammonia channeled there.

The synthesis of urea in mitochondria begins with reaction between carbamoyl phosphate and ornithine. The ammonia produced by the deamination reactions is used to synthesize carbamoyl phosphate.

4.5.1 Synthesis of Carbamoyl phosphate

The first amino group to enter the urea cycle is derived from ammonia in the mitochondrial matrix from deamination reactions. The liver also receives some ammonia via the portal vein from the intestine produced by the bacterial oxidation of amino acids. Such ammonia from above sources is immediately used, together with CO_2 (as HCO_3^-) produced by mitochondrial respiration, to form carbamoyl phosphate in the mitochondrial matrix (Fig. 4.6). This ATP-dependent reaction is catalyzed by carbamoyl phosphate synthatase I, a regulatory enzyme. The carbamoyl phosphate, which functions as an activated carbamoyl group donor, now enters the urea cycle.

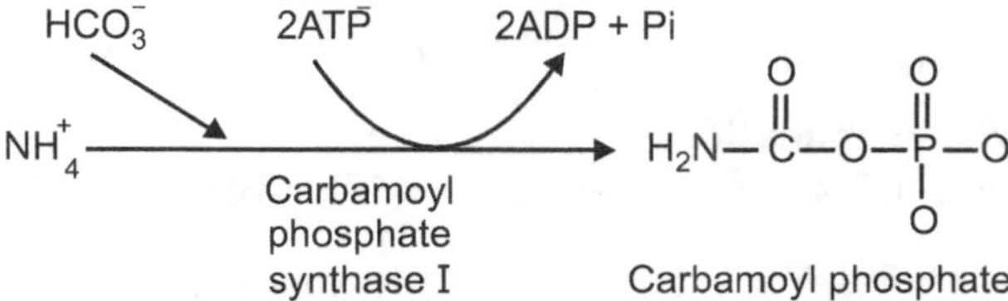

Fig. 4.6 : Sysnthesis of Carboamoyl Phosphate

4.5.2 Enzymatic Steps of Urea cycle

The urea cycle begins inside mitochondria of liver cells, but it is completed in the cytosol; the cycle thus spans two cellular compartments (Fig. 4.7).

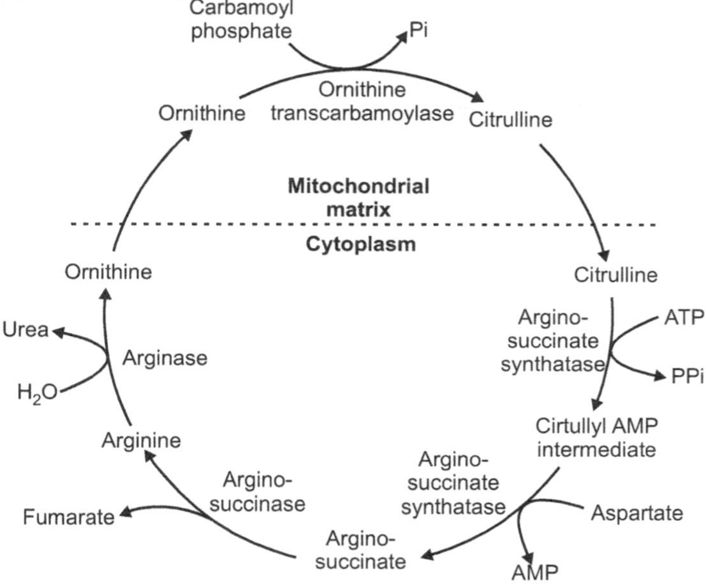

Fig. 4.7 : Urea cycle

4.5.2.1

Step 1 : Donation of Carbamoyl group to Ornithine In first step, carbamoyl phosphate donates its carbamoyl group (containing ammonia from deamination of amino acids) to ornithine to form citrulline, with the release of Pi. The reaction is catalyzed by Ornithine transcarbamoylase, and the citrulline passes from the mitochondrion to the cytosol.

$$H_2N-\overset{\overset{O}{\|}}{C}-O-\overset{\overset{O}{\|}}{\underset{\underset{O^-}{|}}{P}}-O^-$$

Carbamoyl phosphate

$$H_3N^+-(CH_2)_3-\overset{\overset{N^+H_3}{|}}{CH}-COO^- \xrightarrow[\text{trans carbamoylase}]{\text{Ornithine}} H_2N-\overset{\overset{O}{\|}}{C}-NH-(CH_2)_3-\overset{\overset{N^+H_3}{|}}{CH}-COO^-$$

Ornithine   Citrulline

Fig. 4.8 : Donation of Carbamoyl group to Ornithine

4.5.2.2

Step 2 : Condensation of Aspartate and Citrulline The second amino group now enters from aspartate (generated in mitochondria by transamination and transported into the cytosol) by a condensation reaction between the amino group of aspartate and the ureido (carbonyl) group of citrulline, forming argininosuccinate. This cytosolic reaction, catalyzed by argininosuccinate synthatase, requires ATP and proceeds through a citrullyl-AMP intermediate.

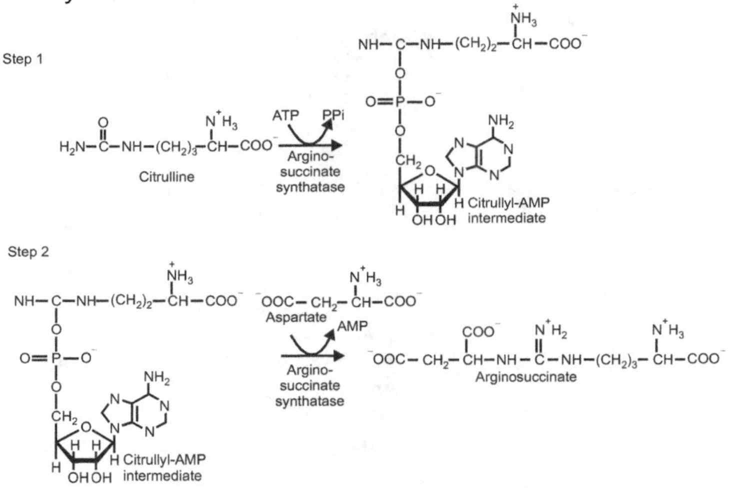

Fig. 4.9 : Condenstion of Asparatate and Citrulline

4.5.2.3

Step 3 : Cleavage of Arginosuccinate The argininosuccinate is then cleaved by argininosuccinase to form free arginine and fumarate. The fumarate enters in the mitochondria to join the pool of citric acid cycle intermediates. This is the only reversible step in the urea cycle.

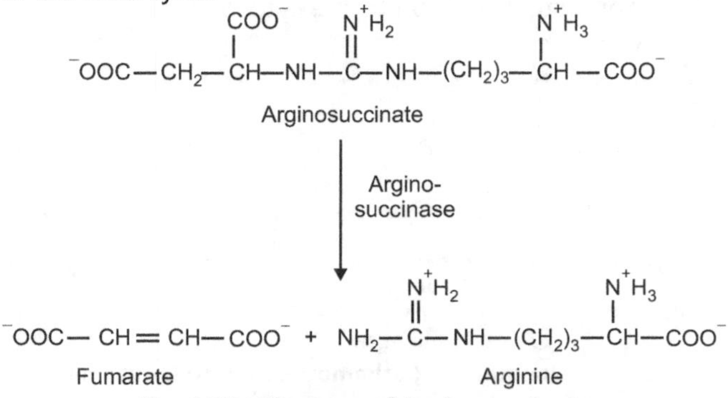

Fig. 4.10 : Cleavage of Arginosuccinate

4.5.2.4

Step 4 : Cleavage of Arginine In the last reaction of the urea cycle, the cytosolic enzyme arginase cleaves arginine to yield urea and ornithine. Ornithine is transported into the mitochondrion to initiate another round of the urea cycle.

Fig. 4.11 : Cleavage of Arginine

The urea produced in the liver cells by the urea cycle is transported through blood to the kidneys. From kidneys it is given out through urine by the mechanism of urine formation.

EXERCISE

1. Write classification of amino acids.
2. Define transamination and explain Transamination reactions.
3. What is Deamination?
4. Explain types of deaminations.
5. Explain in detail Urea Cycle.
6. Write short notes on the following:
    (a) Transamination
    (b) Oxidative deamination
    (c) Synthesis of Carbamoyl phosphate
    (d) Enzymatic Steps of Urea cycle

☆☆☆

5

CHAPTER

ENZYMES

Enzymes are the bimolecules synthesized by living cells of the body. Enzymes are having catalytic activity, hence called **biocatalyst**. The enzymes catalizes biological system and the cascades (chain of reaction). Enzymes accelerate the rate of biochemical reaction without affecting the end product and without undergoing any change, itself in the course of reaction. Thus, "the complex biological catalysts synthesized in living cells which regulate various physiological or biochemical processes of the body are called **enzymes**". The substance on which enzyme acts is called **substrate**.

Nearly all biochemical reactions involving digestion, respiration, stepwise breakdown, synthesis, interconversion of carbohydrates, fats, proteins, nucleic acids and release or utilization of energy are regulated by enzymes. This term enzyme was introduced by **Willy Kuhne** in 1878. The word 'enzyme' is derived from greek and it's meaning is 'in yeast'. The first enzyme was isolated by **Buchner** in 1896, from yeast cells called zymase, a part of enzyme system in fermentation. T. B. Sumner in 1926 isolated enzyme, "Urease" from Jack bean in Crystalline form. For this valuable discovery he received **Nobel Prize** in 1946.

Since then, study of various enzyme resulted in emergence of new branch called enzymology. It has importance in modern Biology, Biotechnology, Genetic engineering, physiology, etc.

5.2 CLASSIFICATION AND NOMENCLATURE

Giving name to the enzyme is called as nomenclature of enzymes. Since 17th century the studies on identification and nomenclature of enzymes have started. Various workers of biochemistry have given the names to the enzymes as per their convenience or thinking. For example,

1. Many enzymes have been named by adding the suffix "-ase" to the name of their substrate or to a word or phrase

describing their activity. Thus, urease catalyzes hydrolysis of urea, and DNA polymerase catalyzes the polymerization of nucleotides to form DNA.

2. Other enzymes were named by their discoverers for a broad function, before the specific reaction catalyzed was known. For example, an enzyme known to act in the digestion of foods was named pepsin, from the Greek *pepsis,* means "digestion," and lysozyme was named for its ability to lyse means to break or destroy bacterial cell walls.

3. Still others were named for their source: trypsin, named in part from the Greek *tryein,* "to wear down," was obtained by rubbing pancreatic tissue with glycerin.

This has caused the ambiguity that the same enzyme has two or more names, or two different enzymes have the same name. Because of such ambiguities, and the ever-increasing number of newly discovered enzymes, biochemists, by international agreement, have adopted a system for naming and classifying enzymes. This system divides enzymes into six classes, each with subclasses, based on the type of reaction catalyzed (Table 5.1). Each enzyme is assigned a four-part classification number and a systematic name, which identifies the reaction it catalyzes. For example, enzyme hexokinase of the first reaction of glycolysis has its name ATP-glucose phosphotransferase and its Enzyme Commission number (E.C. number) is 2.7.1.1. The first number (2) denotes the class name (transferase); the second number (7), the subclass (phosphotransferase); the third number (1), a phosphotransferase with a hydroxyl group as acceptor; and the fourth number (1), D-glucose as the phosphoryl group acceptor.

The International Union of Biochemistry and Molecular Biology (IUB) is involved in the classification and nomenclature of enzymes. The nomenclature committee of IUB has grouped all the enzymes in six classes.

Table 5.1 : International Classification of Enzymes

Sr. No.	Class	Type of reaction catalyzed
1	Oxidoreductases	Transfer of electrons (hydride ions or H atoms)
2	Transferases	Group transfer reactions

Conti..

3	Hydrolases	Hydrolysis reactions (transfer of functional groups to water)
4	Lyases	Addition of groups to double bonds, or formation of double bonds by removal of groups
5	Isomerases	Transfer of groups within molecules to yield isomeric forms
6	Ligases	Formation of COC, COS, COO, and CON bonds by condensation reactions coupled to ATP cleavage

1. Oxidoreductases :

These enzymes catalyses the reaction of oxidation and reduction i.e. removal and addition of H^+ atoms. Hence, these enzymes catalyses the **main energy yielding processes** in the body. These are of two classes i.e. Oxidases and Dehydrogenases.

Dehydrogeanses are the enzymes which catalyse the reaction in which H^+ atoms are removed from one substrate and added / transfered them to the another substrate (except molecular O_2).

$$AH_2 + B \xrightarrow{\text{Dehydrogenase}} A + BH_2$$

The examples of dehydrogenases are Lactate dehydrognease, alcohol dehydrogenase, pyruvate dehydrogenase, Glucose 6-phosphate dehydrogenase, Glutathione reductuse.

Oxidases are the enzymes which catalyse reaction in which H^+ atoms are removed and transfer them to molecular oxygen. In such reaction after reduction H_2O is generated.

$$AH_2 + \frac{1}{2} O_2 \xrightarrow{\text{Oxidase}} A + H_2O$$

e.g. Cytochromo oxidases

2. Transferases :

Transferases are the enzymes which transfer a chemical group from one substrate to another. The transferases are, amino transferases, peptidyl transferases etc.

The amino transferases transfers the amino group from amino acid to the substrate with ketonic structure.

<pre>
          COOH   COOH
           | |
 CH₃   CH₂   CH₃   CH₂
  | | Amino | |
 HC—NH₂ + C=O   ──────────►   C=O   +   HC—NH₂
  | | transferase | |
 COOH   COOH   COOH   COOH
Alanine Oxalo acetic acid   Pyruvic acid Aspartic acid
</pre>

Creatine phosphoryl transferase is the enzyme which transfer energy rich phosphate bond to the ADP and convert it into ATP. Creatine is formed in this reaction.

$$CP + ADP \xrightarrow{\text{CP transferase}} C + ATP$$

Kinases are the enzymes which transfers phosphates group from ATP to the particular substrate. The transfer of phosphate results in formation of phosphate ester.

$$ATP + Glucose \xrightarrow{\text{Kinase}} Glucose\ 6,\ phosphate + ADP$$

e.g. Hexokinase, phosphofructokinase etc.

3. Hydrolases :

Hydrolases are the enzymes which carry out hydrolysis of the substrate by adding water molecule. Water molecule is utilized, for the formation of products.

Examples :

1. Acetylcholine esterase is the enzyme which catalyses the reakdown of a neuro-transmitter acetylcoline into acetic acid and choline. This reaction occurs during transmission of impulse at the synaptic region.

$$CH_3CO - O - CH_2 - CH_2\ N^+ (CH_3)_3 + H_2O \xrightarrow{\text{Acetylcholine esterase}} CH_3 - COOH + OH - CH_2 - CH_2\ N^+ (CH_3)_3$$

     Acetylcholine   Acetic acid   Choline

2. Enzyme sucrose catalyses the breakdown of sucrose into glucose and fructose by the addition of water molecule.

$$C_{12} H_{22} O_{11} + H_2O \xrightarrow{\text{Sucrase}} C_6H_{12}O_6 + C_6H_{12}O_6$$

     Sucrase   Glucose   Fructase

Hydrolases are grouped on the bases of substrates on which they acts e.g. Lipases, Amidases, Phosphotases, Peptidases etc.

4. Lyases :

These are the enzymes which removes the group of atoms from the substrate leading to formation of double bonds or adds group of atoms to the double bonds, without hydrolysis, oxidation or reduction. These acts on C-C, C-O, C-N, C-S and C-halide bonds in the substrate.

e.g. Enzymes fumarases, catalyse the reaction in which malate get converted into fumarate by removal of water molecule

$$Malate \xrightarrow[\text{Fumarase}]{H_2O} Fumarate$$

Decarboxylase is the enzyme of this category which catalyses reaction and remove CO_2.

e.g. Pyruvate decarboxylase

$$CH_3\text{-}CO\text{-}COOH \xrightarrow{\text{Pyruvate decarboxylase}} CH_3\text{-}CHO + CO_2 \uparrow$$

Pyruvic Acid   Acetaldehyde

5. Isomerase or Mutase :

Isomerase is the enzyme which catalyses interconversion of a compound into its isomer. During breakdown of carbohydrate (glucose), Dihydroxy acetone phosphate is converted into 3PGAL (Glyceraldehyde, 3-phosphate) by the enzyme triose phosphate isomerase.

3-Phosphoglyceraldehyde $\underset{\text{isomerase}}{\overset{\text{Triosephosphate}}{\rightleftharpoons}}$ Dihydroxy acetone
(3C)   phosphorus (3C)

(3 PGAL)   (DHAP)

The other isomerases are phosphohexo isomerase, epimarase etc.

6. Ligases or Synthetases :

These are enzymes that catalyse the reaction in which two separate compounds are bind together, on expenditure of energy utilized is usually from ATP by releasing pyrophosphate.

e.g. :

$$X + Y + ATP \xrightarrow{\text{Ligase}} X - Y + AMP + P - P \text{ (Pyrophosphate)}$$

The examples are : DNA ligase, citric synthetase, Glutamine synthetase etc.

5.3 CHARACTERISTICS OF ENZYMES

The characteristics of enzymes are :

(a) Amount or quantity : Enzymes are produced in small amount and required in very less concentration in the reaction. These are not used in the reaction, but available for accelerating rate of reaction. They catalyse large amount of substrate.

(b) Reversibility : Some enzymes shows reversible actions, i.e. interconversion of substrates.

e.g. Triose phosphate isomerase is responsible for,

$$3PGAL \rightleftharpoons DHAP$$

(c) Enzyme Activators : Enzymes may be activated by adding certain activator. In absence of activators enzyme may be inactive or less active.

(d) Enzyme inhibitor : The enzyme activity can be stopped by adding enzyme inhibitor. The inhibitors changes normal condition of the reaction. The salt concentration, metal ion concentration, ultraviolet light and high temperature inhibits the activity of enzymes i.e. they denature the enzymes.

(e) Nature of enzymes : All enzymes are protein in nature except RNAase. The enzymes are proteins with Prosthetic groups and complex structure. They have active sites for binding with the substrate.

(f) Reaction specificity : The enzymes are specific in their reaction. One enzyme usually catalyse one reaction.

e.g. : Oxaloacetic acid shows diverse reaction but, every reaction is catalysed by specific enzyme.

(g) Substrate specificity : Enzymes exhibit substrate specificity, specificity varies from enzyme to enzyme. Substrate specificity is of two types;

(i) Absolute specificity : The absolute specificity is comparatively rare.

e.g. : The enzyme urease shows absolute specificity with its substrate area.

(ii) Relative specificity : The relative specificity is of group and bond specificity. Group specificity is shown by some enzymes where, the enzyme acts on particular groups of different substrates.

e.g. : The enzyme trypsin acts on amino acid, lysine and arginine. The enzyme chemotrypsin acts on aromatic amino acids.

The bond specificity is shown by Proteolytic enzymes. These acts on particular bond in the substrate.

e.g. : Proteases act on glycosidic bond.

(h) Stereochemical specificity : The chemical compound (substrate) shows, different isomers in the living system. The enzymes are very specific with a particular isomer i.e. Substrate.

e.g. : The enzyme succinate dehydrogenase acts on succinic acid and convert it into fumaric acid and not mallic acid which is also isomer of succinic acid.

(i) Rate of reaction : Enzyme accelerates the rate of reaction by lowering energy of activation. Enzyme do not affects the thermodynamics of reaction. The enzyme also do not determine whether the reaction is thermodynamically favorable (exergonic) or thermodynamically unfavorable (endergonic). It has no influence on the amount of reactant involved and product generated in the reaction.

5.4 MECHANISM OF ACTION

The enzyme regulated reactions are carried out at lower energy levels. In such reactions, enzyme accelerates the rate of reaction by lowering energy of activation. In other words, any substrate has to overcome, the lower energy barrier when it get changed or converted into product. The lowering of energy is done by the enzyme by reacting with reactants in a specific way.

The mechanism of enzyme action is explained by various scientists in various ways. Following are some hypothesis about mechanism of action.

(i) Lock and Key model : (Emil Fischer - 1890) :

According to lock and key model, the enzyme has active site which is rigid and not changable. The substrate attaches with active site in a manner like a key fits into a particular lock. Hence, it is called lock and key model.

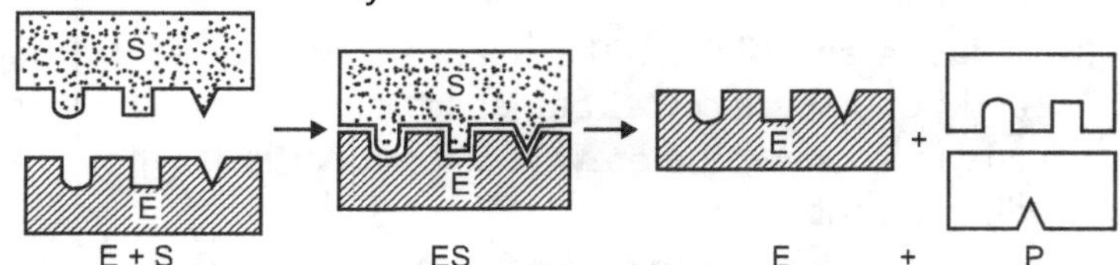

Fig. 5.1 : Lock and key model of enzyme action

(ii) Induced fit model : (Koshland - 1966) :

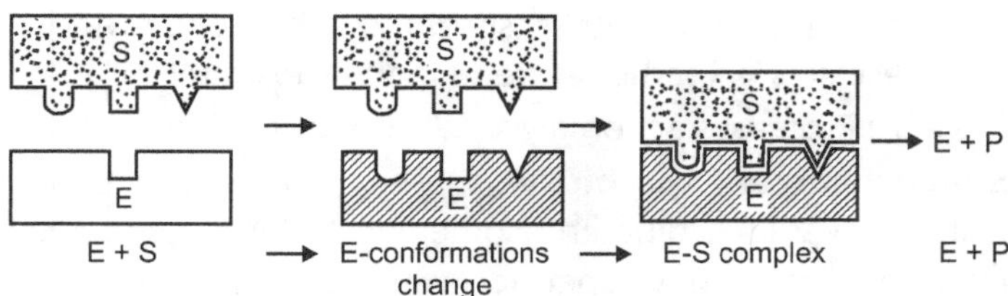

Fig. 5.2 : Induced fit model of enzyme action

According to induced fit model the enzyme does not have rigid structure but undergoes conformational changes to accept the substrate. Due to considerable internal movement conformational changes in the active site occurs to attain final catalytic shape and form. After formation of E-S complex the enzyme acts on substrate and changes it into products.

The induced fit model can be illustrated by the enzyme chemotrypsin that digests proteins by catalyzing the hydrolysis of peptide bonds.

$$\text{Protein} + H_2O \rightarrow \text{Peptide}_1 + \text{Peptide}_2$$

5.5 ENZYME KINETICS

Leonor Michaelis and Maud L. Menten proposed the theory of enzyme kinetics when they were studying effect of enzyme Invertase on hydrolysis of sucrose. This theory is based on following assumptions.

(a) Only a single substrate and a single product are involved.

(b) The process is usually completed.

(c) The concentration of the substrate is much greater than that of the enzyme.

(d) An intermediate enzyme-substrate complex is formed.

(e) The rate of decomposition of the substrate is proportional to the concentration of the enzyme-substrate complex.

The theory postulates that, an enzyme (E) forms weak bond with substrate (S) to form an enzyme-substrate (ES) complex. The ES complex decomposes to yield reaction product (P) and free enzyme (E) by hydrolysis.

$$E + S \longleftrightarrow ES \longrightarrow E + P$$

To describe enzyme kinetics, it is difficult to determine the concentration of ES or S. So Michaelis and Menten used measurable quantities of ES and S. They used following symbols to derive Michaelis-Menten equation.

$[Et] \rightarrow$ Total concentration of Enzyme

$[S] \rightarrow$ Total concentration of substrate

$[ES] \rightarrow$ Concentration of enzyme-substrate complex

$[Et] - [ES] \rightarrow$ Concentration of free enzyme

The rate of formation of products (Velocity - V) is directly proportional to the concentration of the enzyme-substrate complex.

$$V = k \neq [ES] \qquad \ldots 1$$

When total enzyme $[Et]$ binds to substrate, the maximum reaction rate V_{max} will occur. Then the maximum concentration of the $[ES]$ will be equal to the total enzyme concentration $[Et]$.

$$V_{max} = k \times [Et] \qquad \ldots 2$$

Dividing equation 1 by 2, we get

$$\frac{V}{V_{max}} = \frac{[ES]}{[Et]} \qquad \ldots 3$$

For the reversible reaction, $E + S \longleftrightarrow ES$ the equilibrium constant of the ES is K_m which is equal to,

$$K_m = \frac{[Et] - [ES] \times [S]}{[ES]} \qquad \dots 4$$

$$[ES] \times K_m = [Et] \times [S] - [Es] \times [S]$$

$$[ES] \times K_m + [Es] \times [S] = [Et] \times [S]$$

$$[ES] \times (K_m + [S]) = [Et] \times [S]$$

$$\frac{[ES]}{[Et]} = \frac{[S]}{K_m + [S]} \qquad \dots 5$$

Substituting the value from equation 3 to equation 5, we get

$$\frac{V}{V_{max}} = \frac{[S]}{K_m + [S]}$$

$$V = \frac{V_{max} \times [S]}{K_m + [S]} \qquad \dots 6$$

By rearranging equation 6, we get,

$$K_m = [S] \left(\frac{V_{max}}{V} - 1 \right) \qquad \dots 7$$

This is Michaelis-Menten equation to calculate equilibrium constant (K_m) of the enzymatic reaction after determining the reaction rates at various substrate concentrations. It is called as Michaelis constant. It is a measure of the affinity of an enzyme for its substrate. As per equation 4 if the concentration of ES complex is greater, the concentration of free enzyme is lower; consequently the K_m value is lower.

The effect of substrate concentration on the rate of a hypothetical enzymatic reaction, according to the Michaelis–Menten model is shown in Fig. 5.3.

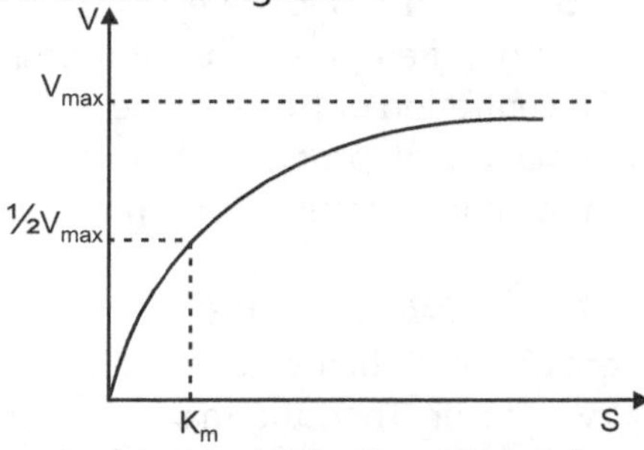

Fig. 5.3 : The Michaelis–Menten plot

Rearrangement of equation 6

$$V = \frac{V_{max} \times [S]}{K_m + [S]} \qquad \dots 6$$

$$V (K_m + [S] = V_{max} \times [S]$$

$$V\,K_m + V\,[S] \;=\; V_{max}\,[S]$$
$$V\,K_m \;=\; V_{max}\,[S] - V\,[S]$$
$$V\,K_m \;=\; [S]\,(V_{max} - V)$$
$$K_m \;=\; \frac{[S]}{V}\,(V_{max} - V)$$
$$K_m \;=\; [S]\left(\frac{V_{max}}{V} - \frac{V}{V}\right)$$
$$K_m \;=\; [S]\left(\frac{V_{max}}{V} - 1\right) \qquad \ldots 7$$

5.6 INHIBITION AND REGULATION

Enzymes are required for most of the processes of life. They catalyze a reaction by reducing the activation energy needed for the reaction to occur. The enzymatic reactions results in the formation of product. However, undesired levels of the product may affect enzyme activity. Hence, enzymes need to be tightly regulated to ensure that product levels do not rise. This is accomplished by enzyme inhibition.

The activity of many enzymes can be inhibited by the binding of specific small molecules and ions. This means of inhibiting enzyme activity serves as a major control mechanism in biological systems. In addition, many drugs and toxic agents act by inhibiting enzymes. Inhibition by particular chemicals can be a source of insight into the mechanism of enzyme action: specific inhibitors can often be used to identify residues critical for catalysis.

5.6.1 Types of Inhibitors

Enzyme inhibitors are chemicals which bind to an enzyme to suppress its activity. They are mainly of two types.

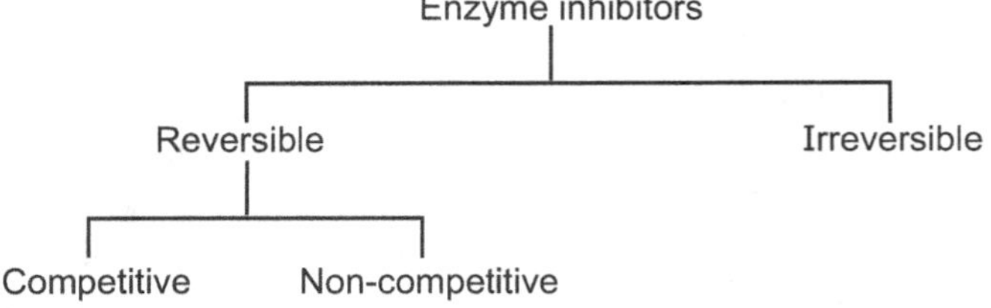

Fig. 5.4

5.6.1.1 Irreversible Inhibitors

The inhibitors which permanently bind to an enzyme are called as irreversible Inhibitors.

5.6.1.2 Reversible Inhibitors

The chemicals which transiently bind to an enzyme are called as **reversible inhibitors.** Reversible inhibitors either bind to an active site (competitive inhibitors), or to another site on the enzyme (non-competitive inhibitors).

5.6.1.2.1 Competitive inhibitors

Competitive inhibitors compete with the substrate at the active site, and therefore increase K_m (the Michaelis-Menten constant). However, V_{max} is unchanged because, with enough substrate concentration, the reaction can still complete. The graph plot of enzyme activity against substrate concentration would be shifted to the right due to the increase of the K_m, whilst the Lineweaver-Burke plot would be steeper when compared with no inhibitor.

5.6.1.2.2 Non-competitive inhibitors

Non-competitive inhibitors bind to another location on the enzyme and as such decrease **V_{MAX}.** However, K_m is unchanged. This is demonstrated by a lower maximum on a graph plotting enzyme activity against substrate concentration and a higher y-intercept on a Lineweaver-Burke plot when compared with no inhibitor.

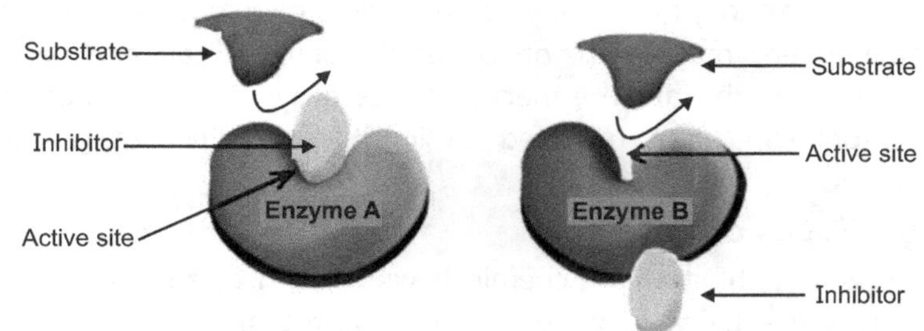

Fig. 5.5 : Competitive and non-competitive inhibitors

5.6.2 Zymogen Inhibition

Enzymes can also be secreted in an inactive state, which are called zymogens. Zymogens are a useful mechanism which allows enzymes to be safely transported to different locations, without the enzyme becoming active and performing its function along the way. They remain inactive due to an addition of amino acids in the protein. Therefore, to activate a zymogen, another enzyme must cleave off these additional amino acids. For example, chymotrypsinogen is synthesized by the pancreas in inactive form.

It is transported to intestines where another enzyme (trypsin) cleaves off the additional amino acids to produce the activated form, chymotrypsin.

5.7 ISOENZYMES, CO-ENZYMES AND CO-FACTORS

5.7.1 Isoenzymes

Isoenzymes or Isozymes are alternative structural forms of the same enzyme that brings about same enzyme action. Thus, isoenzymes are homologous enzymes within a single organism that catalyze the same reaction but differ slightly in structure. The regulatory properties as well as K_m and V_{max} values of isoenzymes are also different. They also differ in electrophoretic mobility. These multiple forms may occur in the same species, in the same tissue, or even in the same cell. Often, isozymes are expressed in a distinct tissue or organelle or at a distinct stage of development.

The different forms of the enzyme generally differ in kinetic or regulatory properties, in the cofactor they use (NADH or NADPH for dehydrogenase isozymes, for example), or in their sub-cellular distribution (soluble or membrane-bound). Isozymes may have similar, but not identical, amino acid sequences, and in many cases they clearly share a common evolutionary origin. The four forms of hexokinase found in mammalian tissues are the best example of isoenzymes.

One of the first enzymes found to have isozymes was lactate dehydrogenase (LDH). In vertebrate tissues, it exists in at least five different isozymes separable by electrophoresis. All LDH isozymes contain four polypeptide chains (each of Mol. Wt. 33,500), each type containing a different ratio of two kinds of polypeptides. The M (for muscle) chain and the H (for heart) chain are encoded by two different genes. In skeletal muscle the predominant isozyme contains four M chains, and in heart the predominant isozyme contains four H chains. Other tissues have some combination of the five possible types of LDH isozymes.

Table 5.2 : Different forms of lactate dehydrogenase

Sr. No.	Type	Composition	Location
1.	LDH_1	HHHH	Heart and erythrocyte
2.	LDH_2	HHHM	Heart and erythrocyte
3.	LDH_3	HHMM	Brain and kidney
4.	LDH_4	HMMM	Skeletal muscle and liver
5.	LDH_5	MMMM	Skeletal muscle and liver

These differences in the isozyme content of tissues can be used to assess the timing and extent of heart damage due to myocardial infarction (heart attack). Damage to the heart tissues results in the release of heart LDH into the blood. Shortly after a heart attack, the blood level of total LDH increases, and there is more LDH_2 than LDH_1. After 12 hours the amounts of LDH_1 and LDH_2 are very similar, and after 24 hours there is more LDH_1 than LDH_2. This switch in the LDH_1/LDH_2 ratio, combined with increased concentrations in the blood of another heart enzyme, creatine kinase, is very strong evidence of a recent myocardial infarction.

5.7.2 Co-enzymes and Co-factors

Enzymes, like other proteins, have molecular weights ranging from about 12,000 to more than 1 million. Some enzymes require no chemical groups for activity other than their amino acid residues. While others require an additional chemical component called a coenzyme and cofactor. Coenzymes are complex organic or metallo-organic molecules while cofactors are one or more inorganic ions, such as Fe^{2+}, Mg^{2+}, Mn^{2+}, or Zn^{2+}. Coenzymes act as transient carriers of specific functional groups. Most are derived from vitamins, organic nutrients required in small amounts in the diet. Some enzymes require both - a coenzyme and one or more metal ions for activity. A coenzyme or metal ion that is very tightly or even covalently bound to the enzyme protein is called a prosthetic group. A complete, catalytically active enzyme together with its bound coenzyme and/or metal ions is called a holoenzyme. The protein part of such an enzyme is called the apoenzyme or apoprotein.

Table 5.3 : Different Cofactors

Sr. No.	Cofactor	Enzyme
1.	Cu^{2+}	Cytochrome oxidase
2.	Fe^{2+} or Fe^{3+}	Cytochrome oxidase, catalase, peroxidase
3.	K^+	Pyruvate kinase
4.	Mg^{2+}	Hexokinase, glucose 6-phosphatase, pyruvate kinase
5.	Mn^{2+}	Arginase, ribonucleotide reductase
6.	Mo	Dinitrogenase
7.	Ni^{2+}	Urease
8.	Se	Glutathione peroxidase
9.	Zn^{2+}	Carbonic anhydrase, alcohol dehydrogenase, carboxypeptidases A and B

Table 5.4 : Different Coenzymes

Sr. No.	Coenzyme	Chemical groups transferred	Dietary precursor in mammals
1.	Biocytin	CO_2	Biotin
2.	Coenzyme A	Acyl groups	Pantothenic acid and other compounds
3.	5' Deoxyadenosylcobalamin (coenzyme B_{12})	H atoms and alkyl groups	Vitamin B_{12}
4.	Flavin adenine dinucleotide (FAD)	Electrons	Riboflavin (vitamin B_2)
5.	Lipoate	Electrons and acyl groups	Not required in diet
6	Nicotinamide adenine dinucleotide (NAD)	Hydride ion $(:H^-)$	Nicotinic acid (niacin)
7.	Pyridoxal phosphate	Amino groups	Pyridoxine (vitamin B_6)
8.	Tetrahydrofolate	One-carbon groups	Folate
9.	Thiamine pyrophosphate	Aldehydes	Thiamine (vitamin B_1)

EXERCISE

1. Define enzymes and give classification and nomenclature.
2. Describe types of enzymes.
3. What are the characteristics of enzyme?
4. Write action mechanisms of enzymes.
5. Explain ENZYME KINETICS of enzymes.
6. Define and explain ISOENZYMES, Co-Enzymes and Co-Factors
7. Write types of inhibitors.
8. Write short notes on the following:
   (a) Lock and Key model
   (b) Induced fit model
   (c) Inhibition and Regulation of enzymes
   (d) Zymogen Inhibition